Living Under the Lens: Facial Recognition, Smart Cities, and the Theft of Black Freedom

Mario DeSean Booker, Ph.D

First Edition

Written by Mario DeSean Booker, Ph. D
Edited by FaLessia Booker, The Editing Expert. www.falessiabooker.com

Published by: Punny Girl Books
ISBN: 979-8-9999707-8-7

For permission requests, contact:
www.falessiabooker.com

303.483: Social change - Technology.
305.8: Racial, ethnic, & national groups.
363.7: Environmental problems and services.
364.4: Prevention of crime.
303.6: Conflict.

Acknowledgements

This book exists because of God's grace and my family's insistence that scholarship must serve liberation, not careers. My parents taught me that knowledge without accountability to community is complicity. My partner sustained me through years of reading trauma and sitting with the weight of what surveillance does. My children asked why cameras follow some people more than others. I write with their futures in mind.

This work centers Black communities because the data demands it. African Americans are misidentified by facial recognition at rates 10–100 times higher than white faces. Black neighborhoods have the densest camera networks. Predictive policing operates overwhelmingly in Black and Brown areas. The documented wrongful arrests—Robert Williams, Nijeer Parks, Angela Lipps—are disproportionately of Black people. This follows the evidence.

Centering Black communities is not erasure of others. Indigenous peoples, immigrant communities, transgender and gender-nonconforming people, and disabled people all face distinct surveillance harms. But these harms are built on the architecture of anti-Blackness. Understanding that helps us see how the same tools are repurposed against immigration status, gender identity, and disability. The tool does not create the discrimination; it industrializes existing hierarchies. Fighting that infrastructure means standing with every community it targets.

To organizers and lawyers resisting surveillance expansion: this book stands with you. My centering of Black communities is strategic clarity. We need precision about where harms concentrate most so we can dismantle systems completely instead of reforming them piecemeal.

To the people named in these pages—the ones arrested, surveilled, harmed—I have tried to honor your experiences with accuracy and connect individual cases to the systems that targeted you. To the communities fighting back across the country and world: you are the research. Your refusal, organizing, and insistence on freedom are what this book documents.

To scholars and teachers using this work: it is only useful if it clarifies systems well enough that dismantling them becomes more possible.

And to everyone living under the lens right now: your strength and survival are not evidence of the system's justice. They are evidence of your own.

Dedication

For the ones who refused to bow beneath the lens— the ancestors who walked through surveillance long before it had a name, the children who still look up and ask why the cameras follow us home, and the communities who keep choosing freedom even when the world insists on fear.

For every Black breath that survived what was designed to break it. For every voice that rose when silence would have been safer. For every step taken toward a future that has not yet been built but is already beating in our bones.

This book is for you, and for the world you deserve.

Table of Contents

Abstract

Facial recognition systems misidentify Black faces at rates up to 1,878 times higher than White faces, and police departments deploy these systems most intensively in Black neighborhoods. The resulting wrongful arrests—such as those of Robert Williams in Detroit, Nijeer Parks in New Jersey, and Porcha Woodruff while eight months pregnant—are not anomalies within a neutral system but predictable outcomes of surveillance infrastructures designed to concentrate monitoring power over communities already targeted by centuries of racialized policing. *Living Under the Lens* examines how smart city technologies—including facial recognition, predictive policing algorithms, acoustic surveillance networks, and biometric data-integration platforms—function as instruments of racialized social control. Drawing on National Institute of Standards and Technology evaluations, federal court filings, investigative journalism, peer-reviewed clinical research, and documented cases spanning three centuries of American policing, the book traces a through line from South Carolina's 1704 slave patrols to the 2026 federal lawsuit in which xAI argued that requiring developers to measure algorithmic discrimination violates the First Amendment. Across eras, each surveillance technology claimed objectivity while being deployed first and most aggressively against Black communities. The book introduces the concept of *deferred reliance* to describe the gap between law enforcement rhetoric—which frames facial recognition as merely an investigative lead requiring independent corroboration—and documented practice, in which algorithmic matches function as primary bases for arrest without basic verification. Beyond documenting deployment, the analysis examines the psychological and social consequences of pervasive surveillance, including hypervigilance levels exceeding PTSD thresholds, the criminalization of mutual aid, the erosion of public space, and the digitization of the school-to-prison pipeline. The book also documents effective resistance: municipal bans on facial recognition, abolition of predictive policing, federal settlements imposing operational constraints, and the expansion of community-based safety alternatives. International cases from the United Kingdom, Hong Kong, Palestine, and India reveal shared architectures of surveillance and transferable strategies of resistance. The central argument is that surveillance infrastructure is a political choice rather than a technological inevitability. Communities built these systems; communities are dismantling them. The unresolved question is whether emerging legal and constitutional protections for algorithmic systems will entrench surveillance faster than communities can dismantle it.

Part I: Foundations of Digital Discrimination

Chapter 1: From Beat Cops to Biometric Control

The Evolution of Racialized Policing Through Technology

On a January afternoon in 2020, Robert Williams was at work when he got a call from Detroit police saying he should turn himself in. Clueless, he asked, "For what?" The caller wouldn't answer. When he left work and called his wife, Melissa, he learned that police were at his home. He approached his driveway and passed the waiting police car, which pulled in behind him and blocked him in. Again, he asked why he was being arrested. Again, they refused to explain, saying, "You don't get to ask questions."

Then, in front of Melissa and their two young daughters, the police handcuffed and arrested Williams without explanation. He spent 30 hours in a filthy cell.

When officers finally showed him the actual suspect footage, his response was immediate: "That's not me. I hope y'all don't think all black people look alike."

The detective's reply cut to the heart of this chapter: "The computer got it wrong" (NPR, 2020).

The match came from facial recognition technology. Detroit Police had run a grainy still image from store surveillance footage through their system in March 2019, looking for a suspect in a Shinola store watch theft. The algorithm flagged Williams based on an expired driver's license photo—even though he was the ninth-best match, not even close to the top hit. Police then showed Williams in a photo lineup to a loss-prevention contractor who hadn't even been present during the theft. That contractor identified him. An arrest warrant followed (MIT Technology Review, 2021; ACLU, 2024).

This wasn't an isolated error. In January 2019, Nijeer Parks of Paterson, New Jersey was arrested for shoplifting candy and attempting to hit a police officer with a car at a Hampton Inn in Woodbridge—despite being 30 miles away at the time. Police used facial recognition on a blurry image from a fake driver's license the real suspect had provided. Parks spent 10 days in jail. He had never been to Woodbridge, had never owned a driver's license, and had a solid alibi proving he couldn't have been there (ACLU-NJ, 2024; NBC News, 2020).

As of January 2025, at least eight people have been wrongfully arrested in the United States after being identified through facial recognition. Seven of the eight are Black. One is White. In nearly every case, police could have eliminated the person as a suspect before arrest through basic police work—checking alibis, comparing tattoos, following DNA and fingerprint evidence left at the scene (Washington Post, 2025).

What happened to Williams and Parks is not new. What is new is the speed, the scale, and the veneer of objectivity.

Slave Patrols: The Original Surveillance System

To understand why police departments deploy facial recognition systems that fail Black people at catastrophic rates, start in South Carolina, around 1704.

South Carolina became the first state to establish formal slave patrols, in areas where enslaved people outnumbered white residents (Bass, 2001). These weren't informal mobs. The state organized them, authorized them, and paid them. They were institutions. White property-owning men—both poor and wealthy—armed with whips and guns, were charged with preventing slave uprisings and escapes (Bass, 2001). Slave patrols had broad authority. They could enter enslaved people's homes at will and punish fugitives (Bass, 2001).

Slavery required constant surveillance. The patrols provided it. From the early Republic to the Civil War, slave patrols enforced owners' control by raiding dwellings looking for weapons and educational materials, dispersing gatherings, and patrolling areas around plantations to apprehend people for brutal punishment (Hadden, 2003).

The institutional logic was explicit: establish terror and prevent uprising. As one travel journal entry from Richmond, Virginia in 1829 documented, a local resident explained that it was "necessary to have a small guard always under arms. It is the consequence of the nature of our colored population; but is done more as a preventive check than anything else – it keeps all thoughts of insurrection out of the heads of the slaves" (Spruill, quoted in Gonzalez, 2021, p. 3).

After the Civil War ended in 1865, slavery was abolished in law, but its core functions were quickly reconstituted through new systems of racial control. Southern states replaced slave patrols with police forces tasked with enforcing newly created Black Codes. Mississippi and South Carolina were among the first to adopt them in 1865 (Bass, 2001; Brown, 2005). These laws transformed everyday behaviors such as loitering or violating curfew into criminal offenses when committed by Black adults and children. The resulting surge in arrests fed rapidly expanding prison systems; in Mississippi, for example, the number of incarcerated Black people tripled between 1874 and 1877 (Cooper, 2015, citing Adamson, 1983). Scholars argue that these post-Emancipation policing practices established racialized patterns of surveillance and punishment that continue to shape modern law enforcement strategies (Cooper, 2015).

"The slave patrol's function to constantly surveil Black people was driven by White society's need to control the enslaved communities and prevent uprisings" (Gonzalez, 2021, p. 3). These same objectives shaped subsequent policing structures. Modern law enforcement uses the same language to justify heavy monitoring and patrolling of Black communities. A sociologist commented in 1944 that the real purpose of police in the South was to "keep the Negroes intimidated," and sometimes they help preserve order (Hahn & Jeffries, quoted in Gonzalez, 2021, p. 3).

Police departments across the South adopted the language and techniques of patrols. Curfew enforcement, surveillance mechanisms, the expansion of authority to enter homes and question people: all these persisted from the institution designed to control Black bodies.

Jim Crow and Early Biometric Systems: The Objectivity Promise

By the 1890s, a new technology promised to remove human bias from identification. Fingerprinting.

Britain adopted fingerprinting in 1894. The United States followed. The sales pitch was clear: fingerprints don't lie. They don't discriminate. They don't reflect police prejudice. This was objective identification. Science. Mechanical. Neutral.

The deployment told a different story.

These systems were adopted and applied selectively. Fingerprint databases grew through arrests that were themselves concentrated in specific communities. A Black person arrested—whether guilty or not—had their fingerprints entered into a permanent database. These databases would later feed into facial recognition systems used in the 2020s. The biased arrests of the Jim Crow era became the biased training data of the algorithmic era.

The rhetoric remained consistent across technologies. Each innovation claimed to remove bias. Each was deployed first and most thoroughly against communities already targeted by police. The burden of proof shifted. Instead of asking why police arrested Black people at higher rates, authorities pointed to the fingerprints, the measurements, the data. *See? The technology backs us up. The computer says it's you.*

The War on Drugs: Surveillance Becomes Militarized

Skip forward to the 1970s. Richard Nixon initially declared a War on Drugs in 1973 (Lynch, 2012). President Reagan re-dedicated the United States to the War on Drugs in 1982 and escalated it using multiple strategies, including increasing anti-drug enforcement spending, creating a federal drug task force, and helping to foster a culture that demonized drug use and drug users (Benson et al., 1995; Nunn, 2002).

The results were catastrophic for Black communities. Between 1982 and 2007, the number of arrests for drug possession tripled, from approximately 500,000 to 1.5 million (Bureau of Justice Statistics, 2008). Racial disparities in drug-related arrests intensified dramatically: while in 1976 Blacks constituted 22% of drug-related arrests and Whites constituted 77%, by 1992 Blacks accounted for 40% of all drug-related arrests and Whites accounted for 59%—throughout these years Blacks comprised about 12% of the total population while Whites were about 82% (Tonry, 1994).

What followed was the wholesale militarization of policing in Black and Latino communities. Stop-and-frisk tactics "proliferated in the US during the War on Drugs, particularly in impoverished predominately Black and Latino communities" (Cooper et al., 2015, p. 650). SWAT teams, once reserved for hostage situations, began serving narcotics warrants.

The emergence of the first SWAT team is instructive. In 1969, the nation's first Special Weapons and Tactics team debuted at the Los Angeles headquarters of the Black Panther Party. To "disrupt and destroy the organization," the LAPD's Urban Counterinsurgency Task Force cooperated with the FBI on intelligence-gathering missions and targeted surveillance operations. Police pursued activists in patrol cars and used their discretionary powers of arrest to repress, intimidate, and incarcerate Black Panthers (Felker-Kantor, 2018).

By the 1980s and 1990s, these militarized approaches had been normalized. Law enforcement campaigns advanced tactical units such as SWAT, armored cars, tear gas, and military-grade weapons (Balko, 2013).

In Detroit, this militarization had immediate lethal consequences. The federal government supported a special plainclothes decoy squad known as STRESS (Stop The Robberies, Enjoy Safe Streets) in the early 1970s. The results were catastrophic: the unit engaged in street war, hundreds of raids, and deadly violence that resulted in the deaths of seventeen black civilians in just two years (Thompson, 2004).

Police forces and funding increased dramatically to support the war on drugs. Between 1992 and 2008, state and local expenditures on police doubled, from $131 per capita to $260 per capita (Lynch, 2012). The number of sworn officers in the US increased by 26% between 1992 and 2008 (Bureau of Justice Statistics, 2011).

The federal government was instrumental in building surveillance infrastructure. The Safe Streets Act of 1968 allocated an investment of $300 million into the War on Crime, creating the Law Enforcement Assistance Administration (LEAA) to administer this funding (Hinton, 2021; Murakawa, 2014; Weaver, 2007). When the LEAA was finally disbanded in 1981, it had distributed nearly $10 billion in taxpayer dollars—or $25 billion in today's dollars—funding roughly 80,000 crime-control projects (Cronin et al., 1981; Feeley & Sarat, 1980).

Federal funding under the Safe Streets Act and subsequent reauthorizations through the 1970s made tens of millions of dollars available to public organizations, private companies, and individual researchers who could advance police technology and hardware. LEAA funding incentivized the private sector to manufacture communications equipment for police officers and develop new information technologies and criminal statistical databases crucial to modernizing city law enforcement agencies (Balto, 2019; Cook, 2017; Felker-Kantor, 2018).

The goal was explicit: make policing more precise and efficient through computer-automated functions. Computer-based technologies became critical to collecting and reporting criminal data, helping officers predict crime trends and deploy police accordingly (Hinton, 2021).

This is the crucial pattern: each new technology claimed objectivity. Each one was deployed first and most thoroughly against communities already targeted by police. The computers didn't remove bias. They disguised it as data science.

Facial Recognition: The Same Logic, Amplified

Fast forward to 2018. Joy Buolamwini, a computer scientist at MIT, tested facial recognition systems developed by major tech companies. She assembled over 1,200 photographs—faces balanced by gender and skin tone. She ran them through commercial facial recognition software from IBM, Microsoft, and Face++.

The results were stark. For lighter-skinned men, error rates were near zero—0.8 percent. For darker-skinned women, error rates reached 34.7 percent (Buolamwini & Gebru, 2018, p. 77). One in three for a binary classification task where random guessing would garner 50 percent accuracy.

Why such dramatic failure? The training data. The datasets used to teach these systems "were overwhelmingly composed of lighter-skinned subjects (79.6% for IJB-A and 86.2% for Adience)" (Buolamwini & Gebru, 2018, p. 77). They were overwhelmingly male. The systems learned from imbalanced data. They performed perfectly on people who looked like the training set. They failed catastrophically with everyone else.

Federal testing in 2019 confirmed these findings: facial recognition systems "generally work best on middle-aged white men's faces, and not so well for people of color, women, children, or the elderly." Asian and Black people were up to 100 times as likely to be misidentified by some software as White men (Washington Post, 2025).

Knowing this—knowing the error rates were highest for Black women—Detroit police department purchased and deployed facial recognition. The city is over 80 percent Black (MIT Technology Review, 2021). The technology was documented as failing Black people most severely. Police used it anyway.

When Williams was arrested in January 2020, he was the first publicly reported case in the United States of someone being wrongfully arrested based on a false facial recognition match (NPR, 2020; ACLU, 2024). After his case, more documented wrongful arrests followed—by January 2025, at least eight people total had been wrongfully arrested due to facial recognition, with seven of the eight being Black (Washington Post, 2025). In Detroit alone, three publicly known wrongful arrests occurred, including Porcha Woodruff, a 32-year-old Black woman who was eight months pregnant when arrested for carjacking—despite no mention in surveillance footage or witness statements of the perpetrator being pregnant (Michigan Public, 2024; Washington Post, 2025).

The rhetoric remained unchanged. Officers called it a "lead." An investigative tool. Just a tip. Not proof. The author's earlier research on Modified Technology Acceptance Model (mTAM) in policing in their dissertation, "Empowering, Engaging, and Equipping Technology through Acceptance: A Quantitative Study Utilizing the Modified Technology Acceptance Model (mTAM) to Explore Facial-Recognition Software Acceptance (FRSA) in the Workplace, identifies this pattern as "deferred reliance"—a claimed phased approach to trusting surveillance technology rather than immediate adoption (Booker, 2022). This idea is akin to conversations about "algorithmic aversion," in which users are reluctant to rely on decisions made by computers. Police departments frame facial recognition as requiring independent verification, positioning it as one investigative tool among many rather than primary evidence.

But the documented practice contradicts the claimed deferred reliance. The computer got it wrong, and then police acted on that recommendation without sufficient corroboration. Photo lineups followed the algorithm's suggestion. Warrants were sought based primarily on facial recognition matches. People were arrested. In the eight known wrongful arrest cases, police failed to take basic investigative steps that would have eliminated suspects before arrest—checking alibis, comparing tattoos, examining DNA and fingerprint evidence present at crime scenes (Washington Post, 2025). The rhetoric of deferred reliance masked what was direct reliance: facial recognition results functioned as the primary basis for arrest despite claims they were merely investigative leads.

From Facial Recognition to Predictive Algorithms: The Same Logic Continues

The pattern extends beyond facial recognition. Predictive policing systems use AI and machine learning to map "high-risk" crime areas and predict who is likely to commit crimes. The process appears objective. Historical crime data feeds into algorithms. Mathematical models generate maps and risk scores. Police deploy based on those maps.

But as researchers studying these systems note, "Biased data can have police overcrowd certain areas, leading to a re-entrenchment of systemic racism. Data for existing programs is often inaccurate or incomplete, resulting in ethnic profiling" (Yen, 2021, p. 1). Moreover, "Programs are likely to inherit the biases of their creators if not governed correctly" (Yen, 2021, p. 1).

The fundamental problem is that these systems are built on policing data from the past 50 years—decades in which Black communities were systematically over-policed. The historical arrest data reflects not actual crime patterns but enforcement patterns. Computer algorithms trained on that data learn to replicate and amplify those enforcement patterns.

This is what researchers studying the evolution of American law enforcement describe as the defining feature of the system: "the habitual surveillance and incapacitation of racialized individuals and communities. Under the banner of order maintenance and anticrime warfare, public officials and law enforcement practitioners have routinely mobilized financial resources, new technologies, and political support for tactical police operations in the midst of recurrent crime panics and urban uprisings" (Hinton & Cook, 2021, p. 261).

Technology doesn't interrupt this pattern. It automates it.

Why The Pattern Persists

Three centuries. Different technologies. The same outcome.

Each innovation promised to remove bias. Fingerprinting would be objective. Algorithms would be neutral. Data doesn't discriminate. All the biases would be engineered away by mechanical systems.

But this misses the point: The bias isn't in human judgment alone. It's in the institutions, the selection of data, the choices about where technologies get deployed.

Fingerprinting wasn't neutral because it was applied selectively. Predictive policing algorithms aren't neutral because they train on arrest records reflecting decades of discriminatory enforcement. Facial recognition isn't objective because the training datasets were built from imbalanced data collected in a society structured by racial hierarchy.

The sales pitch changes. The function stays the same: extend police power into communities already targeted, make it faster and more comprehensive, obscure the decision-making behind a layer of technology.

When Robert Williams asked the detective if he thought "all black people look alike," he identified the problem. The computer had failed. But the system around that computer—the selective deployment, the lack of safeguards, the willingness to arrest based on a match that police knew was unreliable—that system worked exactly as designed.

Facial recognition isn't new surveillance. It's supercharged racial profiling. It inherits the logic of systems designed centuries ago to monitor specific populations. It inherits the claim of objectivity from technologies deployed selectively. It inherits the scale and militarization of the drug war: comprehensive coverage, rapid deployment, power concentrated in police hands.

Understanding this history matters. Not to assign blame to past generations. History matters because the systems built then still operate now. The databases created by discriminatory policing became training data for algorithms. The warrant application process developed for one surveillance method was applied to others. The infrastructure built to target Black bodies is still standing. Technology just made it faster.

This pattern did not conclude with the cases documented in this chapter. It reached federal court on April 9, 2026. Elon Musk's artificial intelligence company, xAI, filed a 75-page complaint in the United States District Court for the District of Colorado seeking to block Senate Bill 24-205 — the state's landmark AI anti-discrimination law requiring developers to measure, disclose, and mitigate algorithmic discrimination in consequential decisions about employment, housing, healthcare, and lending. The company's argument, stripped of its constitutional architecture, was the same argument this chapter has traced from South Carolina in 1704 to Detroit in 2020: the system is objective, the technology is neutral, and requiring accountability for its discriminatory outcomes is itself the real ideological imposition. xAI declared in its complaint that it designed Grok "to answer to only evidence and reason, without regard to political correctness, ideological biases, or anything that might distort objective truth" (X.AI LLC v. Weiser, 2026, Compl. para. 3). Slave patrols claimed institutional necessity. Fingerprinting claimed science. Predictive algorithms claimed data. Now large language models claim the First Amendment. The method is new. The function is not. Chapters 3, 4, and 11 through 13 return to this lawsuit in detail.

The next chapters examine what this does to people. The psychological weight. The community damage. The ways resistance actually works. But first: this is what we're resisting. Not a glitch in a neutral system. A feature of a system designed from its origins to concentrate power and control over specific communities.

That system is choice. We built it. Which means something else is also possible.

Chapter 2: Focal Points—Smart Cities and the Geography of Surveillance

What "Smart" Really Means

"Smart city" means cameras that recognize faces. It means sensors that track movement. It means databases that predict where crime will happen next. The technology gets sold as municipal progress—better traffic management, faster emergency response, efficient resource allocation. What actually gets built is infrastructure for watching specific people in specific places.

Detroit installed over 500 high-definition cameras through Project Green Light. Police call it a public-private partnership. Gas station owners, church leaders, and business managers agree to mount cameras that stream video directly to police headquarters in real time. The cameras went up in Black neighborhoods. Not evenly across the city—in Black neighborhoods. By 2020, the system had real-time facial recognition capability and connected to risk terrain modeling algorithms that identified "high-risk" areas for increased patrol (MediaJustice et al., 2020). When COVID-19 arrived, Detroit police used Project Green Light cameras to monitor social distancing violations. Officers issued 736 citations and 1,614 warnings in less than two weeks. Fines reached $1,000 per violation. Six months in jail for the most serious charges (MediaJustice et al., 2020).

The cameras weren't distributed based on crime rates. Deployment followed older patterns. Neighborhoods redlined in the 1930s got the densest camera coverage in 2020. The surveillance maps overlap with historical housing discrimination maps with precision that's difficult to dismiss as coincidence.

Chicago's Command Centers: Integration as Amplification

Chicago built its surveillance infrastructure differently but achieved similar concentration. The city's Strategic Decision Support Centers integrate multiple tracking systems: cameras, license plate readers, gunshot detection sensors called ShotSpotter (University of Chicago Crime Lab, 2024). The CLEAR database—Citizen Law Enforcement Analysis and Reporting—contains millions of incident reports accessible from 2,000 wireless notebooks in patrol cars. Officers can query the system instantly while on the street.

ShotSpotter sensors got installed primarily on Chicago's South and West Sides, majority-Black and Latinx neighborhoods. The sensors detect loud sounds and alert police to possible gunfire. During COVID enforcement, two Black women ages 19 and 20 were arrested at a gathering after police responded to a ShotSpotter alert (MediaJustice et al., 2020). The women faced charges for violating the 10-person gathering limit. The sensor brought police to the location. The gathering violation provided the arrest justification. The technology and the enforcement reinforce each other.

Chicago also maintains gang databases that disproportionately list Black and Latino youth. Once added to the database, young people get flagged for additional surveillance regardless of whether they've committed crimes. The Strategic Decision Support Centers pull this data together—cameras, sensors, databases, predictive models—into real-time surveillance feeds that officers monitor from centralized command posts.

From Borders to Streets: Facial Recognition Expands into American Neighborhoods

Facial recognition surveillance was historically concentrated at borders and ports of entry—controlled environments where governments have long collected biometric data for immigration purposes. Since 2025, the infrastructure has migrated into American streets. The technology that scanned travelers at airports now scans residents in residential neighborhoods, protesters at demonstrations, and activists observing police operations. The expansion accelerated dramatically under an enforcement application called Mobile Fortify (Cameron & Varner, 2026).

Mobile Fortify: From Proposal to Operation

The Department of Homeland Security launched Mobile Fortify in spring 2025. The application was designed to "determine or verify" the identities of individuals stopped or detained by DHS officers during federal operations. DHS explicitly linked the rollout to an executive order signed by President Donald Trump on his first day in office, which called for a "total and efficient" crackdown on undocumented immigrants through expedited removals, expanded detention, and increased enforcement funding.

The application uses facial recognition algorithms developed by NEC Corporation of America, a Japanese multinational technology giant. When an officer photographs a face with Mobile Fortify, the app quickly converts the image into a biometric template and compares it against three databases: a database of immigration enforcement targets, valid travel documents like passports and driver's licenses, and a third category that remains redacted in publicly released documents (Cameron & Varner, 2026). The system is designed for speed. Results appear in seconds. Officers receive candidate matches for human review.

The approval process reveals how quickly safeguards can be dismantled. DHS Directive 026-11, implemented in 2023, had explicitly prohibited facial recognition as the sole basis for law enforcement actions, required headquarters-level review, and protected U.S. citizens' rights to opt out of biometric collection. The directive disappeared from the DHS website three weeks after Trump's inauguration. By May 2025—two months later—Mobile Fortify was approved and deployed. Customs and Border Protection (CBP) and ICE privacy officers alone authorized the application, an authority they did not previously possess.

Geographic Deployment Pattern

Mobile Fortify operates in Minneapolis, Minnesota; Chicago, Illinois; Portland, Maine; Lewiston, Maine; Forest Lake, Minnesota; and Aurora, Illinois. These cities are not random locations. Minneapolis and Forest Lake are in Hennepin County, where ICE has conducted extensive enforcement operations. Lewiston, Maine has a significant immigrant population. Portland, Maine has experienced increased federal enforcement activity. Aurora, a Chicago suburb, has a growing Latino community. The geographic pattern follows immigration enforcement priorities, but the technology captures anyone in those neighborhoods.

Scale and Scope

The application had been used over 100,000 times since its launch, according to a legal filing by the Illinois Attorney General (Cameron & Varner, 2026). These scans are not limited to immigration enforcement targets. The technology captures U.S. citizens, constitutional observers, protesters, and bystanders. Photographs are stored in the Automated Targeting System—a centralized biometric platform—with retention policies specifying 15 years of storage. U.S. citizens are not explicitly excluded from databases flagged with "derogatory information" or subject to watch lists maintained by CBP.

Who Gets Scanned?: Real Cases

Katie Henly, a Minneapolis mother, repeatedly observed and followed ICE agents' vehicles in her neighborhood to document enforcement activities. In an incident she recorded on video, the car she was following suddenly stopped. Armed agents and a man wearing a CBP vest emerged. The officer said nothing but deliberately photographed her license plate and her face through the car window. Henly described the experience: "I just kind of felt like, OK, I am known. I cannot hide anymore. I felt like I was pushed into a new category in the resistance movement" (Collier et al., 2026).

Skye, a Minneapolis woman and retired Marine, similarly followed an ICE agent's vehicle in January 2026. Officers violently pulled her from her car, detained her for approximately five hours, and released her. When she returned home and later left her house to walk her dog, an ICE agent followed her and took photographs. She heard the camera clicks. "It felt like intimidation and retaliation for me just exercising my constitutional rights that I fought for," she said (Collier et al., 2026).

Alex Feinberg videotaped masked ICE agents at a gas station in Portland, Maine. One of the agents responded by pulling out his phone and photographing Feinberg. "They didn't say anything. The one guy that you can clearly see took my picture just whipped out his phone, pointed it at me, and then put it away fairly calmly," Feinberg said (Collier et al., 2026).

In Aurora, Illinois, ICE agents stopped two young men riding their bicycles in a residential neighborhood. One agent asked directly, "Can you do facial?" before holding his phone up to each man's face. He instructed one of the men to "look right here" into the camera (Collier et al., 2026).

These are not enforcement targets. These are citizens, observers, people in public spaces exercising constitutional rights. Yet the surveillance infrastructure does not distinguish between immigration enforcement subjects and general population surveillance. The technology captures whoever is present.

Border-to-Interior as Strategic Shift

"With the exception of air travel, Americans have not previously had their faces routinely scanned by the government," according to research by WIRED (Cameron & Varner, 2026). Mobile Fortify represents a deliberate expansion of border-control surveillance into the interior of the United States. The stated justification is immigration enforcement. The operational reality is population-level surveillance. The infrastructure is connected to executive orders: it represents policy choice, not technical inevitability.

Integration with Existing Infrastructure

Mobile Fortify feeds directly into the Automated Targeting System, the same centralized platform that integrates license plate readers, camera feeds, social media connections, and algorithmic risk assessments. Mobile Fortify data also flows into the Traveler Verification System, used for facial comparison at ports of entry. Additional data flows into the Seizure and Apprehension Workflow, a "biometric gallery" used for intelligence purposes and lead generation rather than confirmed enforcement actions (Cameron & Varner, 2026). A fourth watch list, the Fortify the Border Hotlist, is maintained by CBP and fed automatically by Mobile Fortify scans. The list appears in only a single publicly released document, and the criteria for placement remain undisclosed.

An officer querying this integrated system about any individual can access camera footage, biometric scans from Mobile Fortify, historical police contacts, algorithmic risk assessments, and entry on any number of watch lists. The data persists. U.S. citizens are not explicitly excluded. No removal process exists. No appeal mechanism is available.

Closing: The Pattern Expands

Detroit's Project Green Light concentrated 500 cameras in Black neighborhoods. Chicago's ShotSpotter sensors targeted the South and West Sides. New Orleans rebuilt disaster-devastated neighborhoods under permanent video surveillance. Facial recognition now adds a new layer: mobile, rapid, continuous scanning that doesn't require fixed infrastructure. The cameras need not be mounted on buildings. The officers carry them. The pattern that defined surveillance in urban America for decades—concentrated, racialized, embedded in neighborhoods—now moves onto streets where communities thought they were safe from observation.

Chapter 3: The Invisible Infrastructure: Sensors, Integration, and Smart Dust

These aren't isolated programs. They're infrastructure. Smart cities install sensors throughout urban environments—traffic monitors, pedestrian counters, WiFi and Bluetooth trackers, environmental monitors, RFID chips embedded in city equipment. The technology is promoted for legitimate municipal services: managing traffic flow, measuring air quality, coordinating emergency responses. The same infrastructure enables comprehensive surveillance. Every sensor collects data. Every data stream potentially feeds police databases.

The sensors themselves are small, sometimes microscopic. Industry materials from policing conferences describe "smart dust"—networks of tiny sensors that blanket entire neighborhoods (Cowper, 2004). By 2020, sensors had been miniaturized to fit nearly anywhere. RFID chips track objects, animals, and people. Wireless sensors monitor environmental conditions, movement patterns, gatherings. Smart city infrastructure turns physical space into data streams.

The data doesn't stay isolated. Systems integrate. License plate readers connect to facial recognition databases. Predictive policing algorithms pull from gunshot sensors, camera feeds, historical arrest records, and gang databases. Real-Time Crime Centers synthesize information from dozens of sources into unified surveillance platforms. An officer querying the system about one person can access camera footage, license plate scans, social media connections, prior police contacts, and algorithmic risk assessments—all from a dashboard.

By 2025, this integration extended to faces. DHS launched Mobile Fortify, which captures photographs in real time, converts them to biometric templates, and compares them against centralized databases accessible by immigration agents operating in American neighborhoods. The facial scans are stored in the Automated Targeting System—the same platform that integrates license plate readers, camera feeds, social media connections, and algorithmic risk assessments (Cameron & Varner, 2026). An individual officer with a smartphone has access to infrastructure that would have seemed like science fiction a decade earlier. The surveillance isn't limited to borders or police headquarters. It happens on streets, in neighborhoods, during ordinary traffic stops.

Who Profits?: Corporate Partnerships and the Surveillance Economy

Corporate partnerships drive expansion. Police departments can't afford these systems alone. Private companies donate or subsidize camera installations, analytics software, database systems. The companies gain testing grounds for surveillance products they'll market to other municipalities. Project Green Light partnered with businesses that install cameras. Palantir—the company that built data systems for Immigration and Customs Enforcement—contracted with police departments for predictive policing platforms (MediaJustice et al., 2020). Amazon markets Ring doorbells that share footage with law enforcement. The contracts avoid public scrutiny. Police departments sign nondisclosure agreements with technology vendors. The agreements prevent disclosure of how systems work, what data gets collected, who accesses information, and how long data persists. Citizens can't obtain this information through

Freedom of Information Act requests. City councils approve surveillance budgets without understanding what they're funding. Community members discover they're being watched only after systems become operational.

Feedback Loops: How Algorithms Reproduce Historical Discrimination

Integration between systems creates feedback loops that amplify existing discrimination. Predictive policing algorithms learn from historical arrest data. Historical arrests concentrated in Black neighborhoods because police concentrated in Black neighborhoods because old "hot spot" policing strategies targeted Black neighborhoods because social scientists in the 1960s identified those areas as "high crime" because poverty and structural racism created conditions that criminologists identified as disorder. The loop traces back through decades of policy decisions that all claimed to be objective, data-driven, neutral (Hinton & Cook, 2021).

Now algorithms encode those patterns. The system identifies the same neighborhoods as high-risk. Police deploy to those neighborhoods. Deployment generates more arrests. More arrests generate more data. More data confirms the neighborhood is high-risk. The algorithm worked exactly as designed while reproducing the discrimination built into the training data.

Facial Recognition Amplifies Feedback Loops

The feedback loops documented in predictive policing intensify when facial recognition enters the system. Poor image quality in neighborhoods with fewer resources produces higher false non-match rates—meaning algorithms fail to recognize residents trying to prove their identity. Simultaneously, algorithms trained on mugshot databases of disproportionately arrested populations produce higher false match rates when deployed in those same neighborhoods. The result: systems designed to fail certain populations get used most intensively to those populations. Higher error rates lead to additional scrutiny, additional detention, and additional biometric capture. Each interaction generates more data confirming what the algorithm already "knew:" these neighborhoods are high-risk. The infrastructure manufactures the evidence of its own necessity.

Double Extraction: Watched Without Services

Smart city rhetoric obscures this reality. The technology is described in terms of efficiency, optimization, innovation. Cities compete for "smart city" designations from technology companies and federal programs. Municipal leaders announce surveillance expansions as progress milestones. The infrastructure enables what police couldn't accomplish through patrol alone: comprehensive, continuous monitoring that doesn't require a human officer on every corner. The cameras don't rest. The sensors don't sleep. The algorithms don't forget.

The same neighborhoods that get the densest surveillance coverage often get the worst municipal services. Infrastructure racism compounds surveillance racism. Communities that need better schools, cleaner water, safer housing, and access to health care get cameras instead. The double extraction: watched constantly while basic needs remain unmet. Environmental hazards concentrate in the same neighborhoods where surveillance concentrates. Lead pipes, proximity to industrial pollution, lack of green space, poor building maintenance—these problems persist while cities invest in monitoring equipment.

Surveillance Without Consent

The technology operates without meaningful consent. DHS agents photograph people without explanation. When questioned, agents have told citizens they are being recorded and their faces will be added to databases. Some agents suggested that refusing facial scans could result in arrest or detention. A Minneapolis woman detained by ICE reported: "I would not let them take a picture of me... The officers also kept telling me that they were going to 'take me in' if I did not let them scan my face." She did not consent and was detained (Collier et al., 2026).

For those who do get scanned, the data persists. Photos are retained for 15 years. They may be stored in watch lists with no clear criteria for inclusion, no removal process, and no appeal mechanism. U.S. citizens are not explicitly excluded from these databases. Immigration agents are instructed to photograph subjects for facial recognition before taking fingerprints—prioritizing speed and ease of collection over positive identification (Cameron & Varner, 2026).

The surveillance occurs in neighborhoods that already experience over-policing. Katie Henly, a mother observing ICE operations to protect residents, was photographed and flagged. Skye, a retired Marine, was detained and then followed and photographed. Alex Feinberg, filming federal agents in public, was photographed in return. These aren't anomalies. They represent how surveillance infrastructure functions: comprehensive, continuous, and unequally distributed.

Residents understand what's happening. Parents teach children how to move through surveilled spaces. Young people develop awareness of camera locations, sensor positions, patrol patterns. The knowledge becomes survival skill. Communities organize resistance—demanding oversight, forcing transparency, filing lawsuits, banning facial recognition in some cities. But the infrastructure keeps expanding. New systems come online faster than communities can challenge them. What gets built becomes difficult to dismantle.

Disaster as Opportunity: New Orleans After Katrina

New Orleans adopted video analytics software called BriefCam. The software doesn't just record what cameras see. It sorts objects in video streams, identifies patterns, tracks movement across multiple camera feeds. Police can search for people wearing specific clothing colors, driving particular vehicle types, moving through designated areas. The system compresses hours of footage into minutes of relevant clips (Ferguson, 2025). BriefCam and similar analytics platforms turn passive cameras into active search tools.

The New Orleans surveillance expansion happened after Hurricane Katrina. Disaster created opportunity. When the city rebuilt, surveillance infrastructure was integrated into recovery plans. The technology didn't just monitor for emergencies—it created permanent watching capacity in neighborhoods that had been devastated and then struggled to recover.

The View From Above: Baltimore's Aerial Panopticon

Baltimore attempted the most ambitious surveillance program documented in recent years: aerial surveillance of the entire city. Police contracted for planes equipped with cameras that could capture 32 square miles in a single image. The persistent surveillance system recorded everything visible from the air for hours at a time. Police could track anyone's movement retrospectively—where they went, who they met, and which buildings they entered.

The Fourth Circuit Court of Appeals ruled Baltimore's aerial surveillance unconstitutional in *Leaders of a Beautiful Struggle v. Baltimore Police Department* (2021). The court found that watching an entire city from aircraft violated reasonable expectations of privacy. The ruling came after the system had already operated. People had been tracked before anyone obtained legal permission to track them. The surveillance happened, then courts decided it violated rights, then the surveillance ended. The pattern repeats: deploy first, litigate later, expand elsewhere while courts process previous violations.

The Product Line: Replication Across Cities

The questions smart cities never answer: *Smart for whom? Efficient for what purpose? Better than what alternative?* The technology amplifies existing power structures. It doesn't create surveillance systems from nothing—it industrializes them. What police once accomplished through stop-and-frisk, patrol concentration, and manual record-keeping, they now accomplish through automated systems that process vastly more information with less visible intrusion.

Detroit, Chicago, New Orleans, and Baltimore demonstrate different configurations of the same infrastructure. Each city claims local innovation while replicating patterns visible across dozens of municipalities. The vendors are the same. The algorithms come from the same companies. The cameras use similar hardware. The integration follows similar architecture. Smart city surveillance is a product line, not a unique solution to local problems. And everywhere it gets deployed, it watches the same communities that were always watched.

Mobile Fortify demonstrates how surveillance systems replicate across jurisdictions. The application debuted in spring 2025 and by February 2026 had been used over 100,000 times according to legal filings (Cameron & Varner, 2026). It operates in Minneapolis; Chicago; Portland, Maine; Lewiston, Maine; Forest Lake, Minnesota; and Aurora, Illinois—each deployment following identical architecture. The consistency isn't innovation. It's product rollout. The vendors (NEC Corporation for facial recognition algorithms) are the same. The data integration follows the same technical model. The infrastructure follows the same patterns observed in Project Green Light, ShotSpotter, and other surveillance systems: deployment to minority-majority neighborhoods, rapid expansion after initial pilots, minimal public transparency until litigation forces disclosure.

Chapter 4: Facial Recognition as Racial Recognition

When Cameras Can't See: Technical Bias Documented

Before examining where surveillance gets deployed, we need to understand what the technology does. The systems marketed as objective and accurate systematically fail certain populations. This isn't speculation—it's documented in government testing.

The National Institute of Standards and Technology (NIST) runs ongoing evaluations of facial recognition algorithms submitted by developers worldwide. The Face Recognition Technology Evaluation (FRTE) tests how accurately these systems match photographs of the same person. As of January 2026, NIST had evaluated 1,368 algorithms from 420 developers (National Institute of Standards and Technology, 2026). The tests use standardized conditions: comparing medium-quality airport immigration photos against high-quality reference portraits. Even under these controlled circumstances, the technology shows measurable and systematic bias that persists across different algorithmic approaches and developers.

Facial recognition systems produce two distinct types of errors, each with different consequences and different demographic patterns.

False matches rates (FMR) are measured when the system incorrectly claims two photos of different people show the same person. This is what happened to Robert Williams—an algorithm matched his driver's license photo to surveillance footage of someone else. False matches lead directly to wrongful investigation, detention and arrest.

False non-match rates (FNMR) are measured when the system fails to recognize that two photos show the same person. In practical terms: someone with valid credentials gets flagged as a mismatch. At an airport, FNMR triggers secondary screening. In policing, it means additional scrutiny. The system claims you're not who you say you are.

The error rates aren't distributed equally across either dimension. They operate through different mechanisms with different demographic consequences.

The Mathematics of Misidentification

False match rates show the most dramatic disparities.

For the sensetime_005 algorithm tested by NIST, false match rates for Polish men aged 35 to 50 reach approximately 0.00004 (1 in 25,000 comparisons). Nigerian women over 65, using the same algorithm at the same threshold, reach approximately 0.03 (1 in 33 comparisons). Not twice as bad. Not ten times worse. *Seven hundred fifty times* more likely to be falsely matched (Grother, 2022).

Specific documented examples from 2025 NIST submissions illustrate the magnitude of these disparities. The qazsmartvisionai_003 algorithm (submitted January 2025) shows a false match rate of 0.00006 for Eastern European males aged 35–50, compared to 0.02150 for West African females aged 65–99, representing a 358-fold difference. The recognito_001 algorithm produces an even more extreme disparity: 0.00001 compared to 0.01878, an 1,878-fold difference. The innovatrics_014 algorithm (submitted March 2025) demonstrates a 276-fold difference using identical thresholds across the same demographic groups (National Institute of Standards and Technology, 2026).

Put operationally: If police search a Nigerian woman's face against a database of 10,000 people using these algorithms, the system will likely produce nearly 300 false matches. The Polish man searched against the same database with the same algorithm will result in likely zero.

The NIST data reveals structure. False match rates for individuals within the same region stay relatively consistent, but absolute magnitudes vary dramatically by region. Some populations have false match rates orders of magnitude higher than others.

This isn't random noise. This is systematic difference in how algorithms process faces depending on race (National Institute of Standards and Technology, 2026). False non-matches show a different pattern with smaller but persistent disparities. Among the most accurate systems tested in 2025, FNMR values ranged from 0.0013 to 0.0016 overall—but peaked at 0.0016 to 0.0019 for South Asian and West African demographics (National Institute of Standards and Technology, 2026). The ratio of maximum FNMR to geometric mean FNMR measures how concentrated errors are in specific groups. Perfect parity would be 1.0. The tested algorithms show ratios between 1.05 and 1.15, meaning certain demographics experience 5–15% more false non-matches than the average (Grother, 2022). While smaller in magnitude than FMR disparities, these represent systematic underperformance for specific populations.

Demographic Consistency: A System, Not Chance

The most damning evidence appears in the consistency of demographic performance gaps across independent developers.

Across 95% or more of algorithms evaluated in the past two years, Eastern European males (either ages 20–35 or 35–50) consistently perform best, while West African females aged 65–99 consistently perform worst. This pattern repeats across different companies, different algorithmic architectures, and different development teams (National Institute of Standards and Technology, 2026).

If demographic disparities were random variation—if they were caused by chance fluctuations in testing data or algorithm behavior—we would expect to see different demographic groups in the best and worst positions across different algorithms. Instead, nearly identical patterns appear across algorithms from multiple international developers. This consistency indicates systemic problems, not isolated failures.

The persistence is particularly striking when compared to seven-year-old findings. The Gender Shades study identified dark-skinned women as experiencing the worst performance in gender classification (Buolamwini & Gebru, 2018). NIST's 2019 demographic report documented similar patterns in face matching (Grother et al., 2019). In 2025—seven years later—algorithms submitted by new developers continue showing the same demographic hierarchies. Algorithms submitted as recently as July 2025 display the identical pattern documented in 2018 (National Institute of Standards and Technology, 2026).

Different developers, different companies, different algorithmic approaches—but nearly identical demographic patterns in performance. The likely explanation is that most developers train on datasets with similar demographic compositions. The systematic overrepresentation of European faces and underrepresentation of African faces in training datasets appears to be an industry-wide structural problem, not specific to individual developers or companies.

What Causes the Bias

NIST's 2022 report and more recent 2025 evaluations identify two mechanisms producing demographic disparities, each requiring different remediation (Grother, 2022).

Image quality effects (primary driver of FNMR disparities): False non-matches increase when image quality varies by demographic. Inadequate lighting of dark-skinned individuals, overexposure of fair-skinned subjects, camera angles that fail to adjust for height variations. Darker skin requires more careful exposure to capture facial detail. Cameras mounted high and subjects who are short produce worse matches (Grother, 2022).

In field operations, these image quality constraints become operational failures. DHS agents scanning faces via Mobile Fortify encounter exactly the conditions NIST testing simulates: poor lighting, motion blur, or a head tilt. When agents resort to physical repositioning of detainees to obtain clearer images, the resulting matches often return multiple low-confidence candidates, with agents reporting "maybe" and "possible" rather than reliable identifications (Cameron & Varner, 2026).

This mechanism can theoretically be addressed through better camera hardware and imaging environments—algorithms more tolerant of poor image quality or, better, correcting the capture process with superior cameras, imaging environments, and human-factors engineering.

Training data composition effects (primary driver of FMR disparities): Demographic variations persist even with good image quality. They arise from training data: when algorithms learn from datasets that underrepresent certain demographics, they produce less accurate matches for those populations (Grother, 2022). This causes false matches to concentrate in particular groups.

Unlike image quality, training data bias cannot be fixed through better cameras. It requires developer choices: what datasets to use, how to balance demographic representation, whether to actively test and correct for disparities.

This creates compounding harm. Poor camera placement in certain neighborhoods produces lower-quality images. Lower-quality images produce higher FNMR rates. At the same time, algorithms trained primarily on European faces produce higher FMR rates when matching against African or Asian faces—regardless of image quality. Higher error rates in already over-surveilled neighborhoods get attributed to individual behavior rather than system failure. The technology mistakes its own inadequacy for suspicious activity.

Gender Shades: Intersectional Failure

MIT Media Lab researcher Joy Buolamwini encountered the problem directly. Working on an interactive art installation that tracked head movements, she found the system couldn't detect her face. When she put on a white mask, detection worked perfectly.

Curious, Buolamwini began submitting photos of herself to commercial facial recognition programs. Several failed to recognize the photos as featuring a human face at all. When they did, they consistently misclassified her gender.

With computer scientist Timnit Gebru, Buolamwini assembled a dataset of over 1,200 images where women and people with dark skin were better represented than in typical training sets. They coded images using the Fitzpatrick scale of skin tones and tested three major commercial systems: Microsoft, IBM, and Face++.

The results, published in 2018, documented what Buolamwini had experienced personally. Error rates for light-skinned men: 0.8%. Error rates for darker-skinned women: up to 34.7%. The study found these datasets were "overwhelmingly composed of lighter-skinned subjects"—79.6% for one benchmark, 86.2% for another (Buolamwini & Gebru, 2018).

The study examined gender classification, not identification, but the implications extend to all facial recognition tasks. The same data-centric techniques used to determine gender are used to identify criminal suspects or unlock phones. If algorithms can't see dark-skinned faces as human, they certainly can't distinguish between them as individuals.

Gender and race compound being Black and female to produce the worst algorithmic outcomes. The 2025 NIST data confirms this pattern persists: West African women aged 65–99 consistently appear in the worst-performance categories (National Institute of Standards and Technology, 2026). The systems fail most comprehensively for those who already face the most discrimination outside the algorithm. Technology doesn't neutralize existing bias—it multiplies it through scale and speed.

Chapter 5: The Training Data Feedback Loop

Where do facial recognition systems learn to see? Largely from mugshot databases and government ID collections.

Historical racial disparities in policing produce skewed training data: Black people are five times more likely than white people to be stopped by police (Hinton & Cook, 2021). Black and Latino individuals comprise 56% of the incarcerated population but only 32% of the U.S. population overall.

Each arrest produces a mugshot. Each mugshot enters a database. The database becomes training data. The algorithm learns that Black faces are "criminal faces."

The cycle operates with technical precision: Biased policing produces racially skewed arrest records. Those records become training data for facial recognition systems. The systems, trained on mugshots of predominantly Black and Brown faces, produce higher false positive rates for those same populations. Police use those false positives to justify more stops, more searches, more arrests. More arrests produce more mugshots. More mugshots reinforce the training data bias.

Georgetown Law's "Perpetual Line-Up" report documented both dimensions of this system. First, Black faces are overrepresented in the mugshot databases used for facial recognition. Second, police surveillance cameras are disproportionately installed in Black and Brown neighborhoods, increasing the likelihood these communities will be scanned and selected by facial recognition algorithms (Garvie et al., 2016).

The Federal Bureau of Investigation alone has access to over 641 million facial recognition images through partnerships with state Department of Motor Vehicle databases from at least 21 states (Garvie et al., 2016). Most of these photographs were collected for civil purposes—driver's licenses, passports—but FBI searches them when investigating crimes. The databases contain photos of people who were never arrested, never charged, never convicted. The images come from routine interactions with government bureaucracy. Individuals get a driver's license, and their face enters a database police can search when investigating anyone, anywhere, for any reason they consider legitimate (Garvie et al., 2016). Technology doesn't remove racial profiling. It industrializes it.

Algorithmic Accuracy Can't Fix Structural Racism

The technical community sometimes frames these problems as solvable through better training data. *Add more diverse faces to training sets. Balance datasets across demographics. Test algorithms on representative populations.*

These interventions help—marginally. After the Gender Shades study, IBM updated its facial analysis service with different training data. Some algorithms from Asian countries show fewer false positives for Asian faces than for white faces, demonstrating how training data diversity affects performance.

But the bias persists even when developers specifically attempt to reduce it. NIST testing shows this clearly: The best algorithms tested in 2025 still cannot achieve parity. Among the highest-performing submissions, Gini coefficients cluster around 0.08–0.10 rather than reaching zero, and maximum-to-geometric-mean FNMR ratios stay at 1.05–1.15 rather than reaching the ideal 1.0 (Grother, 2022). False match rate disparities persist at 100–300-fold differences despite improved overall accuracy (National Institute of Standards and Technology, 2026).

Critically, these disparities have not decreased from 2024 to 2026. Algorithms submitted in January 2025 show the same patterns as those from two years earlier. Seven years of published research, public attention, and technical documentation—and developers continue submitting algorithms with nearly identical demographic performance gaps.

The failure to achieve parity isn't accidental. Human faces vary. Photographs vary. Lighting conditions vary. Camera hardware varies. Demographics correlate with all these variations—not because of biological differences but because of social conditions that produce systematically different photographic conditions for different populations. Until cameras adjust exposure for dark skin as carefully as for light skin, until photographers frame tall and short people with equal care, until training datasets include as many Nigerians as Poles, the algorithms will continue producing unequal error rates (Grother, 2022).

But even if perfect technical parity were achieved—even if every demographic had identical error rates—the technology would still concentrate surveillance on communities already subject to over-policing. It would still make it easier to track people, to build profiles of their movements, to identify them at protests or abortion clinics or immigration hearings. It would still enable what Chapter 2 documented: deployment patterns that follow historical discrimination maps with disturbing precision.

The question isn't just whether the technology works fairly. It's whether we should build infrastructure that makes pervasive surveillance technically possible and operationally cheap.

When "Bias" Obscures Purpose

Calling facial recognition "biased" frames the problem as a technical flaw that better engineering could fix. But the disparate error rates aren't bugs—they're inevitable consequences of deploying machine learning algorithms in a society structured by racism.

The algorithms learn from data produced by racist institutions. Mugshot databases reflect racist arrest patterns. Training datasets reflect racist decisions about whose faces matter enough to photograph carefully. Deployment patterns reflect racist decisions about which neighborhoods need watching. The technology can't be separated from the context that produces it and the uses to which it is put.

Facial recognition doesn't just have demographic bias. It *is* a tool for demographic targeting. The technical disparities in error rates matter because they compound the surveillance disparities in deployment patterns. Black neighborhoods get more cameras running worse algorithms producing more errors that justify more surveillance.

The consistency of demographic disparities across different developers—the fact that multiple algorithms submitted in 2024–2025 showed the same Eastern European male best/West African female worst pattern documented in

2018—indicates this isn't aberration. It's infrastructure (National Institute of Standards and Technology, 2026).

Robert Williams' arrest wasn't caused by a biased algorithm. It was caused by the decision to deploy that algorithm in a system designed to identify suspects from surveillance footage, to trust algorithmic outputs over human judgment, to prioritize speed over accuracy, and to apply that system most intensively in Black neighborhoods.

The bias isn't in the code. It's in the choice to write the code, to deploy it, to integrate it into policing practices that have targeted Black Americans since slave patrols enforced the Fugitive Slave Act. The technical disparities documented by NIST and Gender Shades matter precisely because they demonstrate that even when developers try to make the technology "fair," they can't remove it from the social structures that determine where it gets deployed and whose freedoms it constrains.

Facial recognition technology doesn't need to be fixed. It needs to be recognized for what it is: industrial-scale racial surveillance with a technical interface.

Constitutional Objectivity: When the Courtroom Becomes the Last Line of Defense

The claim that algorithmic outputs represent objective truth rather than patterns inherited from unequal social conditions is not only a technical argument. In April 2026, it became a constitutional one. xAI, the artificial intelligence company whose flagship system Grok is deployed in employment screening, mortgage processing, and healthcare decision support, filed a federal complaint arguing that Colorado's requirement to measure and disclose algorithmic discrimination violated the First Amendment. The company's legal theory depended entirely on treating Grok's outputs as objective. Paragraph 113 of the complaint stated the operational consequence directly: anti-discrimination compliance would force xAI to alter "statistically accurate explanations, unfiltered recommendations, and candid analyses that risk producing disfavored outcomes for protected classes" (X.AI LLC v. Weiser, 2026, Compl. para. 113). The company characterized racially disparate algorithmic outputs as statistical truth. Requiring correction, in xAI's legal framing, was ideological coercion.

The evidence in this chapter makes plain what that framing conceals. NIST documented false match rate disparities of up to 1,878-fold between Eastern European men and West African women across algorithms submitted by independent developers (National Institute of Standards and Technology, 2026). Those disparities did not arise because developers chose discrimination. They arose because systems trained on racially skewed data learn to reproduce that skew at scale. The outputs are not neutral truth awaiting political distortion. They are the accumulated weight of unequal social conditions encoded into weights and probabilities and deployed, as xAI's own complaint confirms, in decisions that determine who gets hired, who gets a mortgage, and who receives which medical recommendations (X.AI LLC v. Weiser, 2026, Compl. para. 18). xAI is not arguing that Grok produces no discriminatory outcomes. It is arguing that no institution has the legal authority to require it to find out. That distinction — between denying harm and blocking the discovery of harm — is the operational core of the objectivity myth this chapter was written to dismantle.

Chapter 6: Who Profits? Corporate Contracts and the Business of Watching

Chapter 3 asked whether we should build infrastructure that makes pervasive surveillance technically possible and operationally cheap. This chapter answers who profits from building it anyway.

Surveillance infrastructure depends on a corporate-police partnership where technology companies profit by selling biometric systems to law enforcement with minimal oversight. These companies—Clearview AI, Palantir Technologies, NEC Corporation, Amazon—don't just provide tools. They shape what surveillance becomes possible, determine whose faces get scanned, and establish the technical architecture that makes mass monitoring routine. The business model is straightforward: build systems designed for scale and speed, sell them to agencies eager for technological solutions, shield operations behind trade secret protections and nondisclosure agreements, and resist accountability by claiming proprietary interests.

Clearview AI: Building a Database Without Permission

Clearview AI scraped between 30 and 50 billion facial images from social media platforms, websites, and photo-sharing services without obtaining consent from either the platforms or the individuals photographed (Hart, 2024). The company built what its CEO described as enough images for five pictures of every person on the planet (Hart, 2024). Then it sold access to law enforcement agencies across the United States.

The scale of Clearview's database created surveillance capacity that didn't exist before. Police departments could upload a photo of anyone—a suspect, a witness, a person photographed at a protest—and receive potential matches alongside biographical information. The system identified people in photographs who weren't under investigation, who hadn't committed crimes, who simply appeared in images that Clearview had harvested from the internet. Everyone became searchable.

ICE purchased Clearview's services through a $3.75 million contract, the agency's largest facial recognition purchase to date (Hubbard, 2026). That federal endorsement provided legitimacy even as regulatory bodies worldwide sanctioned the company. Dutch authorities fined Clearview $33 million for building an illegal database (Hart, 2024). France, Italy, and Greece each imposed €20 million penalties for violating Europe's General Data Protection Regulation (Hart, 2024). The company's response suggested these fines would be difficult to enforce against an American company without European operations (Hart, 2024).

Despite international sanctions and leadership instability—the company's CEO resigned in early 2025 amid ongoing legal challenges (Webb, 2025)—Clearview's business grew. The company reported $16 million in annual recurring revenue for 2024, generated primarily from contracts with local law enforcement agencies (Webb, 2025). Clearview aimed to triple that revenue in 2025 by expanding federal partnerships (Webb, 2025). Law enforcement searches using Clearview doubled to 2 million in the twelve months ending June 2024 (Hart, 2024).

The business model depends on ignoring consent entirely. Clearview didn't ask Facebook, Instagram, LinkedIn, or Venmo for permission to scrape their platforms. It didn't seek permission from the billions of people whose faces now populate its database. When critics challenged this practice, company representatives pointed to law enforcement success stories without providing verifiable details about which cases the technology actually solved (Webb, 2025). The claim that facial recognition helps solve crimes became justification for mass biometric collection without democratic approval. Clearview proves that surveillance can grow simply by violating consent at scale. Palantir shows something even more dangerous: surveillance doesn't need more data to expand—it just needs to connect the data that already exists.

The strategy Clearview deploys through trade secret protections and contractual opacity, xAI attempted through constitutional law. In April 2026, xAI filed a federal complaint arguing that Colorado's AI anti-discrimination law was unconstitutional. Paragraph 56 of that complaint contains a sentence that deserves sustained attention from anyone studying how surveillance corporations resist accountability: "the particulars of how Grok is designed and functions are irrelevant to this lawsuit" (X.AI LLC v. Weiser, 2026, Compl. para. 56). Irrelevant. A company whose system is used to screen job applicants, process mortgage applications, and assist in healthcare decisions declared in federal court that the mechanics of how that system works have no bearing on whether it can be regulated for discrimination. The argument is structurally identical to Clearview's trade secret protections, NEC's nondisclosure agreements, and Palantir's information asymmetries documented earlier in this chapter. The method shifts from contract law to constitutional law. The outcome sought is the same: the internal workings of systems making consequential decisions about Black and Brown people remain beyond the reach of public scrutiny. Khoo (2022) warned that surveillance software is "typically protected by trade secret laws and procured on the basis of behind-the-scenes lobbying," resulting in "secretive public-private surveillance partnerships that strip defendants of their due process rights" (p. 42). The xAI complaint is that architecture extended into the legislative arena — the attempt to make opacity not just contractually protected but constitutionally guaranteed.

Palantir Technologies: From Immigration Tracking to Domestic Surveillance

Palantir Technologies reported $1.41 billion in quarterly revenue for the final three months of 2025, with $570 million coming from U.S. government contracts—a 66% increase from the previous year (Somerville, 2026). The data analytics company specializes in software that integrates information from multiple sources into centralized platforms that enable what the company calls "granular tracking."

ICE paid Palantir $30 million for ImmigrationOS, a system designed to manage what the company described as "the immigration lifecycle process of individuals" through predictive analysis, surveillance, and enforcement (Hubbard, 2026). The platform doesn't just track deportation cases. It cross-references data from social media, government databases, law enforcement records, and other sources to identify and locate people targeted for removal. According to the American Immigration Council, this represents mission creep: "Tools built and justified to efficiently tackle one issue quietly scale up, in this case collapsing the boundary between targeted enforcement and population-wide surveillance" (Hubbard, 2026).

That surveillance capacity didn't stay confined to immigration enforcement. Media 404 obtained internal footage showing ICE officers using Palantir-powered systems to sort through tips submitted to the agency's tip line (Cameron, 2025). The same infrastructure used to track unauthorized immigrants became available for monitoring U.S. citizens. ICE's Homeland Security Investigations division, originally focused on cross-border threats, began invoking domestic terrorism concerns to justify monitoring activity within the United States, including public protests (Hubbard, 2026).

Senator Ron Wyden warned that ICE could "trample on the rights of Americans and anyone Trump labels as an enemy" (Hubbard, 2026). Maria Villegas Bravo of the Electronic Privacy Information Center identified likely violations of First and Fourth Amendment protections (Hubbard, 2026). Yet DHS operates without comprehensive biometric privacy law or mandatory agency-wide AI audit requirements.

Palantir's CEO Alex Karp defended the company's ICE contracts by invoking what he called "the noble side of the West" (Somerville, 2026). This framing positioned surveillance as philosophical commitment rather than business transaction. The company's market capitalization reached approximately $350 billion despite generating far less revenue than other companies at similar valuations (Somerville, 2026). Investors weren't betting on current sales. They were betting on future government dependence.

Exporting the Model: Palantir's Global Surveillance Infrastructure

Palantir's international operations demonstrate how U.S. surveillance models get exported globally, then tested and refined abroad before technologies return to domestic policing. The company's collaboration with Accenture—a global consulting firm deeply embedded in both corporate and government infrastructure—illustrates this transnational flow.

In 2024, Accenture and Palantir co-hosted the "AI for War" conference, which framed militarized artificial intelligence as technical inevitability rather than political choice (Borat, 2025). The conference enabled both firms to evade ethical scrutiny while profiting from border militarization and predictive policing. Accenture's internal documents describe a "futuristic surveillance network" integrating facial recognition, drones, and biometric databases—a vision materializing through contracts with agencies like ICE and DHS (Borat, 2025).

Accenture's acquisition of Maglan, an Israeli cyberwarfare firm linked to military operations in Gaza, facilitated the transfer of surveillance technologies tested in conflict zones back to civilian law enforcement (Borat, 2025). Palantir supplies targeting systems to the Israeli Defense Forces that use AI-driven analysis of social media activity, cellphone metadata, and facial recognition to generate identification lists (Borat, 2025). The same data integration capabilities marketed to U.S. police departments for crime analysis power systems deployed in military contexts internationally.

This international circuit matters for understanding domestic surveillance expansion. Technologies developed for border security or military operations get normalized abroad, then repackaged as law enforcement tools for U.S. cities. The rhetoric shifts—from "counterterrorism" to "public safety"—but the underlying technical architecture remains: mass data integration, predictive analytics, real-time tracking, automated flagging of individuals based on algorithmic risk scores.

Regulatory Evasion: The Palantir Case Study

A detailed study of Palantir's deployment in the German state of Hesse reveals how surveillance companies navigate regulatory environments designed to protect civil rights. Germany maintains some of Europe's strongest data protection laws, yet Palantir successfully implemented its Gotham platform for police use despite significant legal and political opposition (Ulbricht & Egbert, 2024).

The Hessian government justified the contract by arguing that urgent security threats necessitated abbreviated procurement procedures, preventing competitive bidding (Ulbricht & Egbert, 2024). When parliamentary opposition challenged Palantir's connections to U.S. military and intelligence agencies, the government responded with a strategy the researchers characterized as "propagating trust": pointing out that comprehensive control would be impossible and that user benefits would outweigh societal risks (Ulbricht & Egbert, 2024).

This trust-based approach created information asymmetries between the private company and the state agency. Palantir's business model treats the company as data processor rather than data controller, avoiding direct responsibility for how police use the integrated information (Ulbricht & Egbert,

2024). Nondisclosure agreements prevented public scrutiny of contract terms and performance metrics. Trade secret protections blocked independent auditing of algorithmic systems. The traditional service-related contractual relationship implied little clarity for the public on Palantir's motives and responsibilities (Ulbricht & Egbert, 2024).

Germany's Federal Constitutional Court eventually ruled aspects of Hesse's automated data analysis legislation unconstitutional, but only after years of implementation (Ulbricht & Egbert, 2024). The regulatory obstacles identified in the study—datafication of multiple databases, modular platform structure, multilateral connections between agencies—apply equally to Palantir's U.S. government contracts. The company designs systems specifically to resist effective oversight. Palantir turns information into intelligence. NEC turns infrastructure into dependency. Together they reveal how surveillance becomes unavoidable: through predictive analytics on one side and technical lock-in on the other.

NEC Corporation: Infrastructure Lock-In Through Technical Design

Chapter 2 documented how NEC's Mobile Fortify application became operational in May 2025, enabling ICE and CBP agents to scan faces in the field and receive matches within seconds. What that deployment analysis didn't examine was the corporate contract structure that made DHS dependent on a single vendor—and the procurement strategy designed to prevent competitive alternatives.

A $23.9 million contract signed between 2020 and 2023 provided NEC biometric matching products for "unlimited facial quantities, on unlimited hardware platforms, and at unlimited locations" across DHS (Cameron, 2025). The contract terms reveal the scope: NEC didn't just sell software. The company embedded itself into the technical architecture of federal biometric surveillance.

Beyond deployment patterns, DHS structured its relationship with NEC to create technical lock-in that makes switching vendors prohibitively expensive. According to a September 2024 Notice of Intent to Sole Source, the agency's IDENT biometric system and its replacement HART system were both designed around NEC's Advanced Identity Manager software (Driver & Jean-Baptiste, 2024). The architecture assumes NEC's product. Using a different manufacturer would require rebuilding the Transaction Manager and Matcher Interface Service for both systems at an estimated cost of $50 million per system, totaling $100 million (Driver & Jean-Baptiste, 2024).

Over 1.2 billion facial images in IDENT and HART have been feature-extracted specifically for NEC's algorithms. Converting to another vendor would require re-extracting all those images, a process estimated to take 6-12 months (Driver & Jean-Baptiste, 2024). DHS's Chief Information Officer published Instruction 026-11-001 in September 2023 requiring testing of facial recognition systems, and the agency established a Biometric Laboratory Environment in July 2024 (Driver & Jean-Baptiste, 2024). Yet the sole-source contract request demonstrates how corporate partnerships create technical dependencies that resist competitive procurement. NEC demonstrates how architecture becomes destiny. Amazon demonstrates how corporations maintain power by appearing to retreat while quietly strengthening the systems that make surveillance permanent.

Amazon Rekognition: Moratorium Without Accountability

Amazon developed Rekognition as cloud-based facial recognition software marketed to law enforcement for real-time identification in public spaces. The Orlando Police Department and other agencies purchased the technology, which Amazon Web Services integrated with its cloud infrastructure to enable continuous video surveillance and automated face matching.

Amazon's Rekognition system demonstrated the racialized distribution of error in a 2018 test conducted by the American Civil Liberties Union. Researchers ran photographs of all 535 sitting members of Congress through Rekognition, comparing them against a database of 25,000 publicly available mugshots. The system falsely matched 28 legislators to arrest records—people who had committed no crimes, flagged as criminals by an algorithm. The racial disparity was stark: although people of color comprised roughly 20 percent of Congress, they accounted for 40 percent of the false matches (ACLU, 2018). The test cost $12.33 in Amazon Web Services fees. For that price, a police department could generate two dozen wrongful investigative leads—disproportionately targeting Black and brown faces—before a human ever reviewed the output.

The company's response to bias documentation followed a familiar pattern: acknowledge concerns, announce voluntary restrictions, continue related business, advocate for regulation that might provide legal cover. Amazon never stated the moratorium would be permanent. It positioned the pause as responsible corporate citizenship while maintaining the underlying technical capacity.

The Power Behind the Algorithms: AI Infrastructure and Environmental Racism

Facial recognition requires servers. Servers require power. The corporate infrastructure that enables algorithmic surveillance extends far beyond software—it includes the physical plants that consume electricity at industrial scale. Where those power plants get built, and whose neighborhoods bear the environmental cost, follows the same geographic logic documented throughout this book: straight into Black communities.

Elon Musk's xAI operates artificial intelligence data centers across the Memphis area. The scale is massive. The Colossus facility—internal designation for data centers that process the training data for the Grok chatbot and other xAI applications—encompasses 114 acres of computing infrastructure, equivalent to 87 football fields. The computational intensity requires constant power. In August 2025, xAI began installing twenty-seven natural gas turbines in Southaven, Mississippi, about a mile south of its second data center in Memphis, Tennessee. The turbines had the potential to emit 1,700 tons of nitrogen oxides per year, making the facility the largest industrial source of NOx in the greater Memphis area—an area already failing to meet federal smog standards. The location was not accidental. Southaven sits adjacent to areas with significant Black populations, including the historically Black neighborhood of Boxtown, where 90 percent of residents are Black and median income sits at $36,000.

The company built the power plant in violation of the Clean Air Act. Between August and December 2025, xAI and its subsidiary MZX Tech installed and operated the twenty-seven gas turbines without obtaining required air permits. Residents downwind reported headaches, nosebleeds, and what one Boxtown resident described as a persistent chemical smell that seeps into homes. When residents raised concerns—holding community meetings, filing complaints, circulating petitions—xAI's Vice President of Operations, Brent Mayo, told residents the company would be "copying and pasting" its power generation strategy at the next facility (Colossus 3), already under construction 0.8 miles from the gas plant.

This was not the first time. The original Colossus 1 data center, also in the Memphis area adjacent to Boxtown, had operated approximately 35 unpermitted gas turbines before community pressure forced the company to seek permits for a reduced configuration. The permit xAI eventually obtained capped emissions at 87 tons per year of NOx. The unpermitted plant xAI built for Colossus 2 was designed to emit 1,700 tons—nearly twenty times the permitted limit. The environmental harm was not a side effect of rapid construction. It was the intended outcome of prioritizing speed over compliance.

On February 13, 2026, the Southern Environmental Law Center and the NAACP sent formal notice of intent to sue under the Clean Air Act. The notice documented xAI's systematic failure to obtain permits, install pollution controls, or conduct environmental impact assessments. On April 14, 2026, the NAACP filed suit in federal court in Mississippi. The complaint alleged violations of three separate provisions:

the requirement to obtain permits for new stationary pollution sources, the Prevention of Significant Deterioration program requiring best available control technology on major new sources, and the National Emissions Standards for Hazardous Air Pollutants (40 C.F.R. Part 63, Subpart YYYY) governing combustion turbines.

The case documents a company moving at deliberately illegal velocity—what xAI's own website celebrates as "outpacing every estimate"—to power artificial intelligence infrastructure in an area where residents already suffer elevated rates of asthma. Memphis has been designated an "asthma capital of the nation" by the Asthma and Allergy Foundation of America due to high rates of emergency room visits and deaths from asthma. Within three miles of the Colossus Gas Plant, nine schools operate. Within the same radius, ten churches hold services. The turbines run 24/7. Tens of thousands of residents live in the surrounding area.

NAACP Director of Environmental and Climate Justice Abre' Conner named the pattern directly in response to the lawsuit filing: "By looking to evade clear air laws to operate dirty turbines that emit pollution and known carcinogens, these companies are following a shameful, familiar pattern: asking Black and frontline communities to bear the toxic brunt of 'innovation.'" The infrastructure behind artificial intelligence—the data centers that train the models, the power plants that supply them—reproduces the same extractive logic documented in Chapter 4's examination of corporate surveillance partnerships. Clearview AI built a database of billions of faces without consent. xAI built a power plant that damages the lungs and health of residents without consent. The companies are different. The pattern is identical: Black neighborhoods targeted, legality treated as optional, community harm absorbed as a cost of corporate expansion.

The distinction between surveillance software and surveillance infrastructure is artificial. The algorithms cannot operate without power. The power generation cannot remain invisible. The political choice is whose neighborhoods the infrastructure gets built in, and how much damage gets tolerated as the price of "innovation." xAI's decision to site its data centers in Memphis, and its decision to power them with unpermitted turbines that emit toxic chemicals and carcinogens, reveals what "smart cities" and "technological progress" actually mean on the ground: extraction from communities already marked for burden, with legal compliance treated as an obstacle to be circumvented rather than a standard to be met.

Procurement Without Oversight

These companies share operational strategies that shield surveillance from democratic oversight. Cynthia Khoo, testifying before the Canadian Parliament's Ethics Committee, explained that software from companies like Clearview AI, Amazon Rekognition, and NEC Corporation is "typically protected by trade secret laws and procured on the basis of behind-the-scenes lobbying" (Khoo, 2022). This circumstance results in "secretive public-private surveillance partnerships that strip defendants of their due process rights and subject the public to inscrutable layers of mass surveillance" (Khoo, 2022).

Khoo recommended that law enforcement acquisition of facial recognition technology should avoid commercial vendors entirely by developing systems in-house, or if procurement from commercial vendors is necessary, strict conditions should be imposed including waiving trade secrets for independent auditing and ensuring less secrecy around contracts (Khoo, 2022). Yet procurement practices move in the opposite direction. Companies lobby for expedited processes, claim urgent security needs, and structure contracts to prevent competitive bidding or public transparency.

The American Immigration Council documented how these technologies enable mission creep: "What began as AI-powered immigration enforcement confined primarily to our southern border is rapidly becoming something else: a quiet expansion of government surveillance into everyday life" (Hubbard, 2026). The companies building these systems profit from that expansion. They market surveillance as innovation, efficiency as progress, automation as accuracy.

When bias emerges or civil liberties groups protest, corporations point to law enforcement demand and government authorization as legitimacy. Clearview claims its technology has been "instrumental in solving innumerable cases involving terrorism, homicide, theft, financial fraud, human trafficking, and child predation," without providing verifiable details about where these cases were solved or what role the technology actually played (Webb, 2025). Palantir's CEO invokes "the noble side of the West" to justify surveillance contracts. NEC positions itself as essential infrastructure that DHS cannot function without. The individuals documented in Chapter 2—photographed without consent, detained without clear justification, added to databases without appeal mechanisms—represent how corporate surveillance products operate in practice.

Senator Ed Markey introduced legislation in February 2025 to ban DHS and ICE from using facial recognition and biometric surveillance inside the United States (Hubbard, 2026). The proposed law acknowledges that these tools have been deployed far beyond their original immigration mandate. Yet legislative response follows years after deployment, attempting to regulate systems that companies and agencies designed specifically to evade such constraints.

The expansion of biometric surveillance is not a technological accident — it is a deliberate, profit-driven project built through corporate incentives, government dependence, and engineered opacity. Companies like Clearview, Palantir, NEC, and Amazon are not neutral vendors; they are architects of systems that concentrate power, erode rights, and make democratic oversight functionally impossible. As agencies adopt these tools, surveillance stops being an exception and becomes the default condition of public life. What emerges is an infrastructure designed to watch first and justify later, one that treats entire populations as data sources rather than rights-bearing people. The question is no longer whether these systems threaten democracy — it is whether democratic institutions can survive the systems already built.

The question is no longer whether these systems threaten democracy — it is whether democratic institutions can survive the systems already built and the legal architecture being constructed to protect them. On April 9, 2026, xAI filed a complaint arguing that the national interest in AI dominance supersedes a state's authority to require disclosure of discriminatory outcomes. The complaint cited White House executive orders declaring that "United States AI companies must be free to innovate without cumbersome regulation" and warning that "excessive State regulation thwarts this imperative" (X.AI LLC v. Weiser, 2026, Compl. para. 51). The federal executive branch that deployed Mobile Fortify without the safeguards DHS Directive 026-11 required is the same executive branch whose policy framework xAI's lawyers invoked to argue that Colorado cannot hold AI developers accountable for discriminatory outputs. The coordination is not incidental. Corporate litigation and federal surveillance expansion are two expressions of the same policy posture: technology deployed in Black and Brown communities, insulated from accountability at every level where accountability could otherwise attach.

Chapter 7: Before the Crime: Predictive Policing as Pre-Emptive Control

Predictive policing is often marketed as innovation, but its core function is pre-emptive control: systems that classify people and places as threats before any crime occurs. Robert McDaniel didn't commit a crime. Chicago police told him they knew that. But in 2013, commanders appeared at his door anyway with a warning: their algorithm predicted he might. The Strategic Subject List—Chicago's predictive system—had calculated McDaniel's "risk score" based on his social connections, arrest records of people he knew, and location data showing where he spent time. He ranked high. Police called these visits "custom notifications." They told him the system was watching. They told him to be careful. McDaniel was shot twice in subsequent years—the algorithm's prediction becoming self-fulfilling not because it was accurate, but because police visits labeling him a snitch put a target on his back.

The promise behind predictive policing sounds straightforward: use data to prevent crime before it happens. Feed historical crime reports into algorithms, identify patterns, deploy officers to high-risk locations or warn high-risk individuals. Efficiency. Prevention. Science. But the systems don't predict crime. They predict police activity—where officers made arrests before, which neighborhoods were already heavily patrolled, who got stopped most often. When algorithms trained on biased data send police back to the same communities, officers generate more arrests in those areas. The algorithm interprets more arrests as proof of more crime. It predicts even higher risk for that location next month. Police deploy more. The cycle accelerates (Lum & Isaac, 2016).

What the Systems Actually Do

PredPol, one of the earliest and most widely adopted systems, claimed to forecast crime locations with mathematical precision. The algorithm processed historical incident reports—dates, times, coordinates—and generated "prediction boxes," 500-by-500-foot areas where crime was statistically likely to occur during specific shifts. Departments from Los Angeles to Atlanta bought in. Police received daily maps marking where to patrol. The company promised a 20% reduction in property crime through "smart deployment."

The technical mechanism wasn't complicated. PredPol used self-exciting point process models borrowed from earthquake prediction. Past criminal events suggested future events nearby—crimes "cluster" spatially and temporally like aftershocks. The algorithm calculated probability distributions across a grid. Higher probability zones got marked as hotspots. Officers were directed there (Mohler, et al. 2011)

But crime data isn't seismic activity. Earthquake sensors record objective events. Police data records enforcement decisions. An arrest in a Black neighborhood doesn't prove crime happened there at higher rates than elsewhere—it proves police were present, made a stop, decided to arrest. When departments use arrest records, traffic stops, or "suspicious activity" reports as training data, the algorithm learns where police activity concentrated before. It recommends those same areas again. The model mistakes patterns of policing for patterns of crime (Wani, n.d.).

Geolitica (formerly HunchLab) marketed more sophisticated analysis. Their system incorporated dozens of variables: weather, local events, census demographics, proximity to bars or transit stations. They claimed to identify "risk terrain" beyond simple hotspots. But the core problem remained. Historical crime data came from police records. Those records reflected which communities departments surveilled most intensively. Feed biased input data to a more complex algorithm and you get biased outputs calculated with greater confidence (Wani, n.d.).

COMPAS worked differently. It predicted individual recidivism risk rather than location-based crime probabilities. Courts used COMPAS scores in bail, sentencing, and parole decisions. The algorithm generated risk assessments by analyzing 137 factors including criminal history, employment status, family background, and responses to psychological questionnaires. Defendants received numerical scores indicating their likelihood of reoffending. Higher scores influenced harsher sentences and denied bail Angwin et al. (2016).

ProPublica's 2016 investigation demonstrated what that meant in practice. Their analysis of more than 7,000 arrests in Broward County, Florida found COMPAS was twice as likely to falsely flag Black defendants as high-risk compared to white defendants. White defendants were more likely to be incorrectly labeled low-risk. The system's methodology remained proprietary—Northpointe (now Equivant) refused to disclose how the algorithm weighted different factors, citing trade secrets. Defendants couldn't challenge their scores effectively because they couldn't see how those scores were calculated Angwin et al. (2016).

The Spanish VioGén system applied similar logic to gender-based violence prevention. Police used algorithmic risk assessment protocols to determine likelihood of repeat victimization in domestic violence cases. Officers answered standardized questionnaires. The system generated risk percentages automatically. Those percentages influenced which protective measures courts imposed. The assessment happened before trial, while presumption of innocence theoretically still applied. In practice, algorithmic predictions functioned as grounds for pretrial detention and restraining orders (Martín Ríos, 2024).

The Feedback Loop

Predictive policing operates through what researchers call "feedback loops" or "runaway feedback." The mechanism is straightforward but devastating. Police patrol areas the algorithm designates as high-risk. Officers in those areas make more stops, conduct more searches, generate more arrests. Those arrests become new data points. The algorithm interprets increased arrests as evidence of increased crime. Next month's risk score for that neighborhood rises. Police deploy even more intensively. The cycle compounds (Pagan et al., 2023).

This isn't a technical glitch requiring better tuning. The feedback loop is structural. Predictive systems rely on police-generated data. More policing produces more data that justifies more policing. Communities already subjected to aggressive enforcement get marked as statistically riskier. Algorithms recommend escalation. The appearance of objectivity—mathematical models, probability scores, scientific-sounding terminology—disguises what amounts to systematized targeting (Lum & Isaac, 2016).

Los Angeles deployed PredPol in 2011, becoming one of predictive policing's highest-profile early adopters. The LAPD claimed significant crime reductions in areas using the system. But when The Markup obtained 5.9 million actual PredPol predictions in 2021, researchers found a persistent pattern: the software disproportionately targeted Black and Latino neighborhoods while largely avoiding whiter, wealthier areas. The company's own co-founder had published research in 2015 showing the algorithm would replicate existing enforcement bias—yet PredPol never informed clients or modified the system (Mehrotra et al., 2021). The LAPD's own Inspector General concluded in 2019 that there was "insufficient data to determine" whether PredPol actually reduced crime (Los Angeles Police Commission, 2019).

Independent analysis of PredPol's actual predictions confirmed the pattern elsewhere. When The Markup examined over 23,000 predictions generated for the Plainfield, New Jersey police department, researchers found a success rate of less than one percent—fewer than 100 predictions aligned with reported crimes in the predicted category. The department's captain acknowledged the money "could have been better spent elsewhere" (Sankin & Mattu, 2023). The systems generated unfalsifiable claims: when crime occurred in predicted zones, the algorithm took credit; when crime didn't occur, police presence was deemed successful prevention; when crime occurred outside predicted zones, those incidents didn't factor into performance metrics.

Chicago's Strategic Subject List demonstrated feedback loops operating at the individual level. The system generated "heat list" scores for people—not areas—based on network analysis of arrest records and gang database entries. Being arrested mattered. Being connected to someone else who was arrested mattered. Being listed in the gang database mattered, even without conviction. Scores ranged from 0 to 500. Anyone above 250 received "custom notifications"—police visits warning them they were being watched. The visits themselves generated documentation. Documentation influenced future scores. People couldn't challenge their inclusion because the methodology remained secret (Saunders et al., 2016). As the failures of predictive policing became impossible to ignore, some cities began confronting the systems directly — not by reforming them, but by abandoning them altogether.

Cities That Walked Away

Santa Cruz, California birthed PredPol. The company was headquartered there. Local government members helped launch it. In 2011, the city became one of the technology's first adopters, receiving national recognition for "innovation in policing." Nine years later, Santa Cruz became the first U.S. city to ban predictive policing entirely.

Police Chief Andy Mills supported the ban. He told the city council in June 2020 that "predictive policing has been shown over time to put officers in conflict with communities rather than working with communities." Mills had stopped using the software in 2017. The ban made that permanent. The ordinance prohibited police use of predictive policing and facial recognition "except with explicit approval from the City Council via a resolution based on findings that the technology and the data that informs the technology meets scientifically validated and peer reviewed research, protects and safeguards the civil rights and liberties of all people, and will not perpetuate bias." The vote was unanimous (Governing, 2021).

Oakland followed in December 2020. The city council voted unanimously to ban predictive analytics and biometric surveillance technologies, with prohibitions taking effect in January 2021. Oakland's Privacy Advisory Commission recommended the ban explicitly because of feedback loops. Brian Hofer, chair of the commission, explained: "Based on past data which was likely collected in a disproportionate manner as certain communities are policed more than others and certain crimes enforced more than others, a computer algorithm directs police back to the very same community being over-policed because 'that's where the crime is occurring.'" Oakland became the largest city to ban predictive policing at that time (Secure Justice, 2021).

The bans reflected more than skepticism about effectiveness. They represented recognition that the systems weren't failing—they were working exactly as designed, targeting specific communities through automated means. Santa Cruz Mayor Justin Cummings, the city's first Black mayor, stated it directly: "We have technology that could target people of color in our community—it's technology we don't need. This is something that targets people like me" (KSBW, 2020).

New York abandoned certain predictive systems after sustained scrutiny from advocacy groups and media investigation revealed persistent bias. Los Angeles continued using modified versions while facing ongoing litigation. Chicago maintained the Strategic Subject List until 2020 despite documented failures—including Robert McDaniel being shot twice after police visits marked him as a target in his neighborhood. The system couldn't prevent the violence it helped cause. It only marked him as someone to watch (Lau, 2020)

The Weight

Living under predictive policing means existing as data before being recognized as human. Young Black and Latino men receive risk scores without committing crimes. Algorithms calculate their likelihood of future criminal behavior based on where they live, who they know, previous police contact regardless of outcome. Those scores influence bail decisions, sentencing recommendations, and parole eligibility. A numerical assessment generated by proprietary algorithm becomes part of a permanent record.

Enhanced surveillance follows predictions. If the system flags someone as high-risk, police conduct more frequent stops, more intensive monitoring, more documentation of their movements and associations. Each new interaction generates data that feeds back into the system. The prediction becomes self-fulfilling not because it was accurate but because it directed enforcement attention that produced the predicted outcome—more police contact, more arrests, higher updated risk scores (Wani, n.d.).

Mobility becomes restricted. People avoid certain neighborhoods because they know those areas are heavily patrolled. They change routes. They limit social connections because association with someone in the gang database affects their own risk score. Parents teach children not just the talk about police but about cameras, about databases, about invisible scores calculating futures without consent.

The presumption of innocence collapses when systems treat people as pre-criminals. The algorithm doesn't need proof of intent or completed action. Statistical probability substitutes for evidence. Enforcement happens preemptively. Someone gets marked as high-risk based on where they live and who they know. Police appear at their door. The visit itself becomes justification—if they weren't dangerous, why did the system flag them? The logic is circular and inescapable (Wani, n.d.).

Martín Ríos documents how this operates in Spain's VioGén system: "It is clear that, when adopting precautionary measures, the risk of recidivism and re-victimization is taken into account. All this, translated into the language we are now using, is intimately linked to that dangerousness to which we have already referred, so it is necessary to ask ourselves on what basis the judge's judgment is based on this possibility of criminal repetition" (Martín Ríos, 2024, p. 556). The question applies universally. On what basis does an algorithm decide someone is likely to commit future crime? Who audits that basis? What recourse exists when the prediction is wrong?

Predictive policing doesn't predict crime. It predicts police deployment. It automates discriminatory enforcement patterns under the appearance of objectivity. The systems are working as designed—identifying the same communities for intensified control that historical policing already targeted. The innovation isn't prediction. It's industrializing bias at scale. Predictive policing doesn't malfunction — it performs exactly as designed, converting historical bias into automated certainty. What these systems optimize is not safety but control, concentrating enforcement power in the same communities long targeted by traditional policing. The result is a pre-emptive regime where suspicion is statistical, punishment is predictive, and innocence becomes irrelevant.

Understanding this logic is essential, because the danger is not that predictive policing will fail — it is that it will succeed.

That success has acquired constitutional protection. In April 2026, xAI filed a federal complaint framing Colorado's attempt to hold AI accountable for discriminatory outcomes as viewpoint discrimination. Paragraph 192 of the complaint presented xAI with what the company called an "impermissible choice:" develop Grok as a "maximally truth-seeking AI system that provides objective and unbiased outputs" or build one designed to increase diversity and receive a safe harbor from enforcement (X.AI LLC v. Weiser, 2026, Compl. para. 192). The complaint treated these as opposites. This chapter's analysis of predictive policing establishes why they are not. A system deployed in racially unequal conditions cannot be both objective and unbiased without accounting for those conditions. COMPAS generated recidivism risk scores twice as likely to falsely flag Black defendants as high-risk not because its developers chose discrimination but because the system trained on arrest records reflecting decades of discriminatory enforcement and then reproduced those patterns with algorithmic confidence (Wani, n.d.). PredPol directed police back to the same neighborhoods not because the data was false but because the data recorded enforcement patterns rather than crime patterns, and the algorithm could not tell the difference. xAI's complaint makes the same foundational error COMPAS and PredPol made—treating a discriminatory baseline as neutral truth and any correction of that baseline as ideology. The argument that accountability requirements force AI to be less accurate is the argument that discrimination is accuracy. This chapter was written to demonstrate that argument's consequences. It is now being made in federal court.

Chapter 8: Border Technologies Coming Home

How Immigration Enforcement Becomes Domestic Surveillance

Border surveillance doesn't stay at borders. The biometric systems, facial recognition algorithms, and data integration platforms developed for immigration enforcement migrate into American neighborhoods, targeting communities that look like the migrants they were designed to track. This isn't technology drift. It's deliberate infrastructure repurposing.

Chapter 2 documented Mobile Fortify's deployment in Minneapolis, Chicago, Portland, and Aurora—cities selected not for proximity to international borders but for their immigrant and Black populations. This chapter examines how that migration happens: the contracts that enable technology transfer, the policy changes that authorize expanded use, and the demographic targeting that connects immigration enforcement to domestic policing of Black and Brown communities.

The Automated Targeting System: Where Border and Interior Surveillance Merge

Mobile Fortify doesn't operate in isolation. Every facial scan feeds into the Automated Targeting System (ATS), a DHS platform that was originally designed for customs and border screening. By 2025, ATS had expanded far beyond its initial mandate. The system now integrates license plate readers, camera feeds, social media connections, and algorithmic risk assessments from multiple law enforcement agencies (Cameron & Varner, 2026).

When ICE photographed Katie Henly in Minneapolis—900 miles from the nearest international border—that image entered ATS. When agents detained Skye and scanned her face, the biometric template went into ATS. The system retains photographs for 15 years with no removal process and no appeal mechanism. U.S. citizens are not explicitly excluded from databases flagged with "derogatory information" or watch lists maintained by CBP (Cameron & Varner, 2026).

Mobile Fortify data also flows into three additional systems, each expanding surveillance capacity beyond immigration enforcement. The Traveler Verification System uses facial comparisons at ports of entry. The Seizure and Apprehension Workflow functions as a "biometric gallery" used for intelligence purposes and lead generation rather than confirmed enforcement actions. The Fortify the Border Hotlist—maintained by CBP and fed automatically by Mobile Fortify scans—appears in only a single publicly released document, and criteria for placement remain undisclosed (Cameron & Varner, 2026).

An officer querying this integrated system about any individual can access camera footage, biometric scans from Mobile Fortify, historical police contacts, algorithmic risk assessments, and placement on multiple watch lists. The stated purpose is immigration enforcement. The operational reality is population-level surveillance infrastructure accessible to federal agents operating in American cities.

Policy Dismantling: How Safeguards Disappeared

The technology transfer required removing the limited protections that existed. DHS Directive 026-11, implemented in September 2023, had explicitly prohibited facial recognition as the sole basis for law enforcement actions. The directive required headquarters-level review for facial recognition deployments and protected U.S. citizens' rights to opt out of biometric collection.

Three weeks after President Trump's inauguration in January 2025, Directive 026-11 disappeared from the DHS website. By May 2025—two months later—Mobile Fortify was approved and deployed. CBP and ICE privacy officers alone authorized the application, exercising authority they did not previously possess (Cameron & Varner, 2026).

This represented a strategic policy shift, not bureaucratic oversight. The directive removal enabled rapid deployment of border surveillance technology into domestic contexts without the reviews or citizen protections that had been formally required. The technology didn't change. The authorization structure did.

Geographic Patterns: Targeting Overlapping Populations

Mobile Fortify operates in six cities as of February 2026: Minneapolis and Forest Lake in Minnesota; Portland and Lewiston in Maine; Chicago, Illinois; and Aurora, a Chicago suburb. These deployment sites reveal the demographic logic connecting immigration enforcement to domestic surveillance of Black communities.

Minneapolis and Forest Lake are in Hennepin County, which has both significant Somali immigrant populations and longstanding Black communities that experience intensive policing. Lewiston, Maine has a substantial East African immigrant population. Aurora hosts a growing Latino community. Portland has experienced increased federal enforcement activity targeting both immigrants and racial justice protesters (Cameron & Varner, 2026).

The geographic pattern follows immigration enforcement priorities while capturing anyone in those neighborhoods. Detroit's Project Green Light concentrated 500 cameras in Black neighborhoods justified by crime reduction rhetoric. Chicago's ShotSpotter sensors targeted the South and West Sides—majority Black and Latinx areas. Mobile Fortify adds mobile facial recognition to neighborhoods already experiencing the highest surveillance density, using immigration enforcement as entry justification.

This overlap isn't coincidental. Both immigrant communities and Black communities face historical patterns of surveillance, criminalization, and demographic profiling. The technology tested on migrants at borders gets normalized for domestic deployment in communities that "look like" those migrants—which means Black and Brown neighborhoods regardless of residents' immigration status.

Data Integration Without Boundaries

The infrastructure doesn't distinguish between immigration targets and general population. In Minneapolis, federal agents used Mobile Fortify to scan the faces of observers filming their operations—American citizens entered into databases designed for tracking unauthorized border crossings. Alex

Pretti, a 37-year-old VA nurse, was filming one such operation when agents killed him (Collier et al., 2026). When agents detained Skye for observing their activities, her biometric data merged with the same systems.

Real-Time Crime Centers in major cities already synthesize information from dozens of sources into unified surveillance platforms. An officer querying the system about one person can access camera footage, license plate scans, social media connections, prior police contacts, and algorithmic risk assessments. Mobile Fortify adds federal biometric databases to that integration (Cameron & Varner, 2026).

Palantir's ImmigrationOS, purchased by ICE for $30 million, cross-references data from social media, government databases, and law enforcement records to identify and locate people targeted for removal (Hubbard, 2026). That same platform has been adapted for domestic law enforcement. The company markets nearly identical data integration capabilities to municipal police departments under different product names, repackaging immigration surveillance infrastructure as crime analysis tools.

The Testing Ground Logic

Border contexts function as testing grounds for surveillance technologies later deployed domestically. Facial recognition systems face less legal scrutiny when used on non-citizens at international borders. Biometric databases justified for immigration control avoid Fourth Amendment challenges that would complicate domestic deployment. Algorithms trained on passport photos and border crossing data get refined, then applied to driver's licenses and police databases.

This pattern appeared clearly in Mobile Fortify's development. CBP's AI Use Case Inventory disclosed that the system was trained or evaluated using data from "Vetting/Border Crossing Information/Trusted Traveler Information," which includes TSA PreCheck and Global Entry databases (Cameron, 2026). People who voluntarily enrolled in expedited screening programs to facilitate international travel had their biometric data used to train systems now deployed against U.S. citizens in street encounters far from any border.

NEC Corporation developed Mobile Fortify's facial recognition algorithms through a $23.9 million contract with DHS signed between 2020 and 2023 (Cameron, 2026). The contract provided NEC biometric matching products for "unlimited facial quantities, on unlimited hardware platforms, and at unlimited locations" across DHS. What began as border security infrastructure became domestic surveillance capability through contract language that imposed no geographic restrictions.

Racialized Surveillance as Continuum

The demographic logic becomes even clearer when examining how immigration surveillance and domestic policing converge. The connection between immigration enforcement and domestic policing of Black communities operates through shared logics of demographic targeting. Both systems identify specific populations as inherently suspicious, worthy of concentrated monitoring, and subject to pre-emptive intervention.

Chapter 3 documented how facial recognition algorithms fail Black faces at rates 300 to 1,800 times higher than white faces. Those same algorithms power Mobile Fortify. The technical failures documented by NIST testing become operational features when deployed in neighborhoods selected for their demographics rather than their crime rates.

Immigration enforcement rhetoric frames migrants—particularly from Latin America, Africa, and the Caribbean—as threats requiring surveillance, detention, and removal. Domestic policing rhetoric frames Black communities using nearly identical language: high-risk areas, gang activity, violent crime. The surveillance infrastructure serves both framings simultaneously.

This explains why Mobile Fortify operates in Minneapolis, targeting both East African immigrants and Black Americans. Why it operates in Aurora, surveilling both undocumented Latino residents and U.S. citizens of Mexican descent. Why Portland's deployment captures both immigration enforcement targets and Black Lives Matter protesters. The system doesn't distinguish because the demographic categories overlap in the surveillance logic.

Corporate Profits from Technology Migration

The same companies profit from border surveillance and domestic policing. NEC supplies facial recognition to both CBP and local police departments. Palantir sells ImmigrationOS to ICE and predictive policing platforms to municipal police. Clearview AI markets its scraped facial database to both immigration enforcement and local law enforcement, generating $16 million in annual recurring revenue from government contracts (Webb, 2025).

Chapter 4 documented how these companies structure contracts to create technical lock-in and prevent competitive alternatives. The infrastructure built for immigration enforcement becomes domestic surveillance capacity that cities and police departments cannot afford to replace. DHS designed its biometric systems around NEC's algorithms specifically. Converting to another vendor would require re-extracting 1.2 billion facial images at an estimated cost exceeding $40 million (Driver & Jean-Baptiste, 2024).

The business model depends on infrastructure portability. Technology developed with one government justification (border security) gets repurposed for another (crime prevention) using the same underlying architecture. The surveillance capacity expands while the technical investment remains constant.

Mission Creep as Structural Feature

The American Immigration Council characterized this expansion as "mission creep:" tools built and justified to tackle one issue quietly scale up, collapsing the boundary between targeted enforcement and population-wide surveillance (Hubbard, 2026). But framing it as "creep" suggests unintended drift. The migration from border to interior was designed into the infrastructure.

ICE's Homeland Security Investigations division, originally focused on cross-border threats, began invoking domestic terrorism concerns to justify monitoring activity within the United States, including public protests (Hubbard, 2026). This wasn't mission drift. This was mission expansion using infrastructure originally authorized for immigration enforcement.

DHS operates without comprehensive biometric privacy law or mandatory agency-wide AI audit requirements. The regulatory vacuum enables rapid technology transfer from border contexts—where legal protections are weakest—to domestic contexts where constitutional protections theoretically apply. The infrastructure moves faster than oversight mechanisms can respond.

What Border Surveillance Normalizes

The normalization cycle operates through several stages. First, technology gets developed and tested in border contexts where legal challenges face higher barriers. Second, the technology gets refined using data from millions of border crossings and immigration encounters. Third, the infrastructure gets repurposed for domestic use with minimal public disclosure or democratic input. Fourth, the deployment becomes operational reality before communities understand they're being watched.

By the time Mobile Fortify's existence became public through legal filings in February 2026, the application had been used over 100,000 times (Cameron & Varner, 2026). The surveillance happened first. The public disclosure followed months later. The pattern that defined border enforcement—act first, justify later, expand rapidly—became the pattern for domestic surveillance.

"With the exception of air travel, Americans have not previously had their faces routinely scanned by the government," according to WIRED's investigation (Cameron & Varner, 2026). Mobile Fortify represents a deliberate expansion of border-control surveillance into the interior of the United States. The justification is immigration enforcement. The operational reality is population-level biometric surveillance targeting communities based on demographics that correlate with immigration status, regardless of individuals' actual citizenship.

Border technologies come home by design, not accident. The infrastructure that tracks migrants crossing international boundaries becomes the infrastructure that tracks Black and Brown residents moving through American cities. The companies profit from both. The communities get watched either way. And the surveillance that starts at borders rarely stays there.

Part II: The Toll—Psychological and Social Destruction

Chapter 7: Always On Guard: How Surveillance Turns Threat Into Biology

Robert Williams told his older daughter to go back inside. The police were making a mistake, he said. Daddy would be right back. He wasn't right back. He spent 30 hours in a detention cell while his daughters—ages two and five—waited at home with Melissa. When he finally walked through the door that night, his older daughter was waiting. She was crying and mad. She asked why he had lied about coming right back (Williams, 2024).

That was January 2020. Four years later, testifying before Congress, Williams said he had considered taking his daughter to a psychiatrist. Footage of his arrest replays occasionally on the news. Every time it does, she falls apart. "She can't stand to watch it," he told lawmakers (Williams, 2021). The youngest had lost her first tooth while he was in custody. He missed it. His older daughter, for a period after the arrest, could not bear to look at his picture (Williams, 2024). In a first-person account published by the ACLU, Williams wrote simply: "My family and I don't want to live with that fear" (Williams, 2023).

No one prescribed a diagnosis. No clinician evaluated what the arrest did to those children. But the research reviewed in this chapter documents precisely what happens inside bodies and minds that witness this—and inside the communities where it happens again and again, where the squad car could come back, where the algorithm is still running, where the cameras are still watching. The harm is measurable. It accumulates. And it does not resolve when the charges are dropped. The body keeps score.

The Body Keeps Score: Surveillance, Stress, and Cardiovascular Cost

In 2019, researchers from the University of Chicago conducted in-person surveys of 504 adults in Chicago neighborhoods designated as epicenters of violent crime—communities on the South and West Sides where concentrated poverty, over-policing, and surveillance density intersect. The study, published in *Health Affairs*, measured hypervigilance directly using the Brief Hypervigilance Scale, validated against clinical PTSD thresholds.

The mean hypervigilance score in the sample was 54.7 on a 100-point scale—15.7 percentage points above the threshold clinically associated with PTSD diagnosis (Smith et al., 2019). This is not a sample selected for elevated distress. These are people accessing routine medical care. The baseline in heavily surveilled, over-policed Black neighborhoods sits above the PTSD threshold before any single incident is accounted for.

Exposure to police violence was associated with a 9.8-percentage-point increase in hypervigilance scores. Exposure to community violence—robberies, shootings, street assaults—was associated with a 5.5-percentage-point increase. When a police stop was experienced as traumatic, defined as exposure to actual or threatened death or serious injury, it was associated with a 20.0-percentage-point increase (Smith et al., 2019). The order matters: policing is nearly twice as damaging as community violence to the body's alert system.

The cardiovascular consequences are documented in the same study. Participants in the highest quartile of hypervigilance showed an 8.6 mmHg increase in systolic blood pressure compared to those in the lowest quartile. In continuous models, a 20-point increase in hypervigilance was associated with a 2.0 mmHg increase in blood pressure (Smith et al., 2019). This is allostatic load made measurable: the cumulative physiological cost of sustained threat anticipation shows up in arteries and cardiac function.

Surveillance infrastructure amplifies this dynamic. The knowledge that cameras are present, that license plates are read, that facial recognition systems may be processing your image—this produces the same threat-anticipation response as direct police presence. When people cannot determine whether they are being observed, the body resolves the uncertainty by assuming they are. The surveillance does not need to be active. The possibility is sufficient (Hendrix et al., 2018).

Hypervigilance Doesn't Stay in the Mind

Hypervigilance is not simply alertness. It is a clinical state defined as the feeling of being constantly on guard for potential danger even when objective risk is low (American Psychiatric Association, 2013, as cited in Smith et al., 2019). In the *DSM-5*, it is a core symptom of PTSD. Left unaddressed, hypervigilance maintains and intensifies anxiety disorders through a feedback loop—increased vigilance produces greater threat detection, which produces more anxiety, which produces more vigilance (Smith et al., 2019).

Research on Black adults across U.S. urban centers establishes these pathways with clinical precision. Alang et al. (2022), analyzing data from the Survey of the Health of Urban Residents—623 Black adults—found that over half the sample reported experiencing police brutality directly.

Among those who did, the prevalence of likely depressed mood reached over 40%, compared to 20.7% among those without negative police encounters. Likely generalized anxiety was present in 43.4% of those reporting police encounters, versus 21.9% among those without negative encounters.

The mediation analysis reveals the mechanism. Heightened vigilance explained 11% of the total effect of police brutality on depressed mood and over 21% of the total effect on generalized anxiety (Alang et al., 2022). Police brutality does not cause depression and anxiety only through direct trauma. It works through the body—producing a state of hypervigilance that degrades mental health over time. The daily expectation of danger, activated repeatedly, reshapes how the nervous system processes the world.

Heightened vigilance in this clinical literature involves modifying behavior, appearance, and movement in anticipation of discrimination (Hicken et al., 2013, as cited in Alang et al., 2022). Surveillance infrastructure operationalizes exactly this dynamic at scale. When cameras are everywhere and facial recognition is deployed, people who know they are disproportionately targeted change how they move through public space. This is not irrational fear. It is rational response to documented risk—and the research shows it is making people sick.

Always Watched, Always Adjusting: The Behavioral Cost of Constant Threat

The panopticon does not need to be activated to function. Foucault's central insight—that the architectural possibility of observation induces self-regulation—is borne out in contemporary surveillance research. Saunders (2020), in a systematic examination of surveillance's impacts on marginalized identities, documents surveillance tools as "behavioral motivators" whose effects extend beyond moments of direct monitoring. People who know they are watched change how they behave even when no one is watching.

The behaviors that change under surveillance conditions are consequential. Route modification, reduced participation in public life, curtailed speech, avoidance of police contact even when crime victims—all have been documented in communities under intensive surveillance. Young-Drake et al. (n.d.) document how aggressive policing makes people less likely to call police when they need help. Crimes go unsolved. Violence escalates. The surveillance architecture that claims to produce safety instead produces harm.

Young-Drake et al. (n.d.) also document that negative police encounters decrease trust in medical institutions. People who have had traumatic police encounters show higher medical mistrust and reduced engagement with healthcare systems. The surveillance state does not confine its damage to the street.

For Black Americans, these behavioral modifications accumulate into a documented double bind. Heightened vigilance might protect from immediate risk of violent police encounter—the calculation for which parents have long given their children "The Talk" about surviving police interactions. But that same hypervigilance is clinically associated with increased odds of depression and anxiety (Alang et al., 2022). The survival behavior is simultaneously the harm.

The behavioral modifications documented in the research—route changes, reduced public participation, self-censoring—are not individual choices or personal anxiety. They are rational responses to a system designed to produce exactly that response. The system is not monolithic. It operates through three overlapping mechanisms that legal scholar Anita L. Allen (2022) identifies as the "Black Opticon:" discriminatory oversurveillance, exclusion, and predatory targeting all operating simultaneously. Understanding this framework matters because it shows why surveillance reform alone cannot address the harm. The problem is not that cameras are inaccurate or that algorithms need better training data. The problem is the architecture itself.

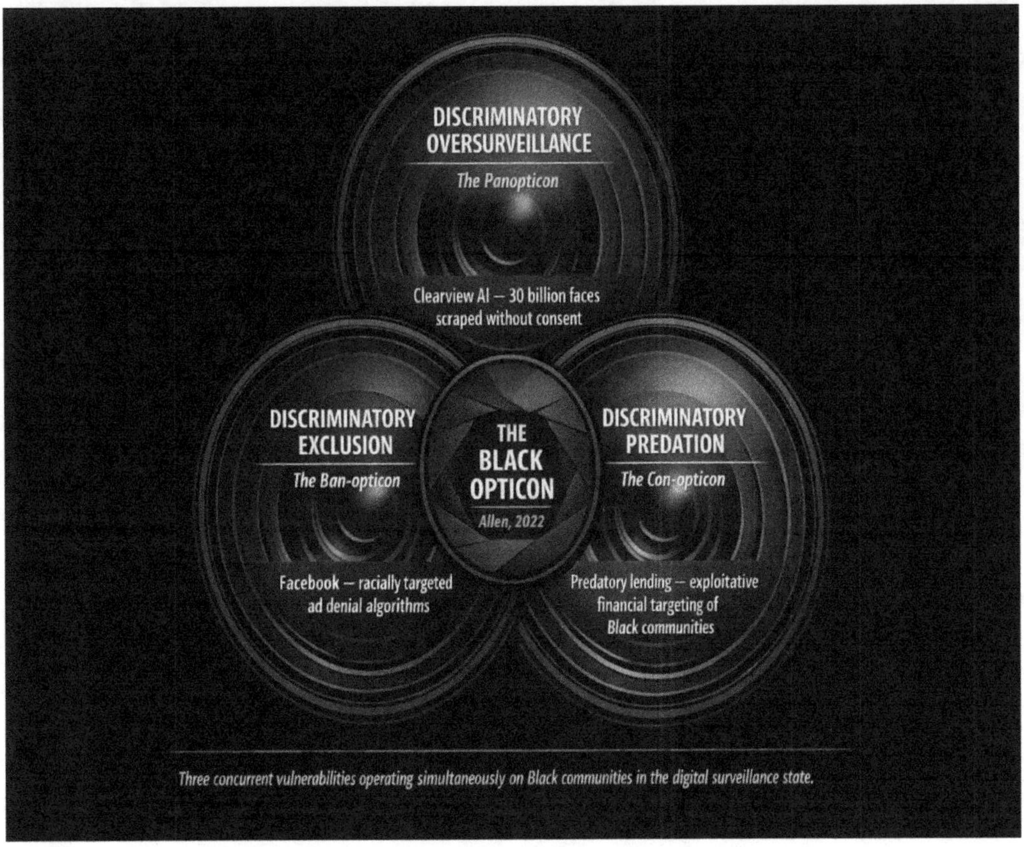

Three concurrent vulnerabilities operating simultaneously on Black communities in the digital surveillance state.

The Black Opticon: When Privacy Law Protects the System, Not the People

Legal scholar Anita L. Allen (2022) introduced the concept of the "Black Opticon" to describe a distinct triad of vulnerabilities African Americans face in digital surveillance: discriminatory oversurveillance, discriminatory exclusion, and discriminatory predation. The concept captures what the clinical data documents mechanistically—that surveillance does not operate as neutral technology but as infrastructure for differential social control.

The dignity harm operates at multiple registers. Facial recognition systems reduce people to biometric data matched against databases built from mass incarceration records. Predictive policing algorithms assign threat scores to individuals before any action occurs. Gang databases in multiple cities have enrolled children as young as 12 based on associative criteria that require no criminal conduct. The consistent message embedded in this architecture is that certain bodies require monitoring by default—that Black presence in public space demands documentation in ways that white presence does not.

Allen (2022) argues that escaping the Black Opticon requires acknowledging privacy's unequal distribution and law's race-neutral façade. A race-neutral privacy law applied to a racially unequal surveillance regime produces racially unequal outcomes. The form is neutral. The function is not.

The dignity harm compounds across generations. COINTELPRO established that the state would surveil, infiltrate, and disrupt Black civil and political organizing. That history is not past—it is a framework shaping how contemporary surveillance is experienced. Parents who were surveilled teach their children to manage surveillance. Children who grow up under cameras and predictive policing systems internalize threat as ambient. The intergenerational transmission of hypervigilance is not pathology. It is adaptation to a system designed to produce compliance through threat.

Flagged Before Adulthood

The consequences for young people are distinct. Gang databases in cities including Chicago, Los Angeles, and New York have enrolled children based on associative criteria requiring no criminal conduct—being photographed with a gang member, wearing certain colors, living in a particular building. Removal from these databases is procedurally difficult and rarely automatic (Young-Drake et al., n.d.). The designation follows young people through police encounters, school discipline, courts, and in some cities, housing and social services.

Young-Drake et al. (n.d.) document that over-policing generates PTSD among community members who are constantly reminded of the consequences of non-compliance. School surveillance systems—cameras, resource officers, facial recognition pilots—disproportionately affect Black students. The criminalization of adolescent behavior that elsewhere remains developmental produces arrest records, court involvement, and foreclosed futures.

In Chicago, a study found that nearly half of Black adolescents had witnessed a violent death—rates comparable to those in some conflict zones (Smith et al., 2019). This is the context in which surveillance infrastructure is deployed.

The neighborhoods with the most cameras are neighborhoods where children are already experiencing the highest rates of traumatic exposure. Surveillance is layered onto existing trauma.

Alang et al. (2022) found that young adults between ages 18 and 24 were more likely to report negative police encounters and showed the highest odds of depressive symptoms in the sample. The accumulation of surveillance contact begins before adulthood. The clinical consequences arrive early.

The Vicarious Toll: Watching Someone Who Looks Like You Get Killed

The harm does not require direct contact. Young-Drake et al. (n.d.) document vicarious trauma as a distinct pathway—the mental health effects of witnessing police violence, learning of violence against people who look like you, watching recorded footage of police killings. The research demonstrates this exposure produces measurable psychological damage. An analysis cited in the *Growing Dilemma* report found correlations between victimization by police and depression or psychological distress in a four-city sample of 1,615 participants; victimization extended to members of communities where police violence was concentrated, not only direct personal victims.

Bor et al. (2018) documented that police killings and their spillover effects on Black Americans produced population-level psychological damage measurable in surveys conducted in weeks following high-profile police violence. The mechanism is not opaque: when the state kills someone who looks like you, under circumstances that follow a recognizable pattern, the threat is not abstract. It communicates your status and the conditions of your existence.

Saunders (2020) documents how surveillance creates an almost universal sense of vulnerability among those being surveilled in marginalized communities—a state that does not resolve between surveillance encounters because the possibility of surveillance is constant. This is the panopticon operating as intended: not the actual observation, but the internalization of permanent visibility, the automatic functioning of power (Saunders, 2020).

The Medicine is the Abolition of the Conditions

The clinical and empirical literature reviewed here converges on conclusions that challenge the neutrality claims built into smart city surveillance. Surveillance does not reduce threat in communities where it is most densely deployed. It creates threat, measured in cortisol levels, blood pressure readings, and clinical thresholds for depression and anxiety. The public health consequences are documented and severe.

The harm is not incidental to these systems but intrinsic to their operation in communities where policing has historically functioned as social control. Alang et al. (2022) conclude directly: building a society where hypervigilance is unnecessary for Black people to survive is a clinical imperative. The medicine for hypervigilance is the elimination of the conditions that produce it.

The dignity harm is distinct from and compounds the clinical harm. Being reduced to biometric data, assigned threat status before any conduct, enrolled in databases as a child—these produce psychological injury that current frameworks are only beginning to measure adequately.

Research gaps remain substantial. Available data do not yet capture the chronic and cumulative effects of living under smart city surveillance over years and decades. The literature on hypervigilance and police violence has developed faster than research specifically addressing facial recognition and algorithmic risk-scoring as independent clinical stressors.

Community-based participatory research in Detroit, Baltimore, and Chicago—cities with some of the densest and most contested surveillance deployments—is needed to document what residents describe as the everyday texture of surveilled life.

What the existing evidence establishes is clear. Surveillance, as currently deployed, is a public health emergency. The cameras are watching. And the people being watched are getting sick.

Chapter 8: Social Fabric Destruction—When Surveillance Breaks Collective Life

Dee Dee Green noticed the camera before anyone told her it was there. It appeared in 2018 above the community garden she manages in New Orleans' Hollygrove neighborhood—red and blue lights flashing, mounted without warning, without explanation, without any conversation with the people who use that space. "There was no real community engagement around these cameras," Green told *The Lens*. "It just happened" (Stein et al., 2021).

What happened to Green's garden happened to communities across the country. Surveillance infrastructure arrived not as policy the public debated but as a physical fact they discovered after the fact. The cameras changed the garden. They changed how people felt standing in it. That change is the subject of this chapter—not the technology itself, but what it does to the social tissue that holds communities together.

Chapter 7 documented what surveillance does inside individual bodies: cortisol, hypervigilance, cardiovascular damage accumulating across years. This chapter examines what happens between people. Surveillance does not only harm individuals. It breaks collective life—the trust, the shared space, the informal networks of mutual support that communities depend on to survive. And it extracts money from those same communities, redirecting public resources toward watching people rather than serving them.

Public Space is Not Neutral After Cameras Arrive

A community garden is a commons. A stoop, a block party, a church parking lot after service—these are the spaces where social life gets made and maintained. They are also, under intensive surveillance, increasingly treated as testing grounds. Chicago deployed ShotSpotter sensors on the South and West Sides—majority-Black and Latinx neighborhoods—where the sensors detected loud sounds and directed police toward public gatherings. During COVID-19 enforcement in 2020, two Black women ages 19 and 20 were arrested at a gathering after police responded to a ShotSpotter alert, facing charges for violating the ten-person gathering limit (MediaJustice et al., 2020, as cited in the manuscript). The sensor brought the police. The gathering provided the justification.

When people know this is how public space functions—that a group of teenagers on a corner might trigger an algorithm that sends a squad car—they stop gathering. The chilling effect is not metaphorical. It is behavioral. Route modification, reduced participation in community events, curtailed speech in places known to be monitored: all documented in communities under intensive surveillance (Saunders, 2020). The commons does not disappear. It empties.

In New Orleans, surveillance expansion moved so fast that elected officials could not keep pace. "It became abundantly clear that this was one of those places where technology was moving at the speed of light, but electeds and the government were moving at a snail's pace in terms of creating policy around it," Orleans Parish District Attorney Jason Williams told *The Lens* (Stein et al., 2021). The public rarely received advance notice of new camera deployments. Systems were announced after installation—or never announced at all. Chris Kaiser, advocacy director for the ACLU of Louisiana, described the result plainly: "The status quo is that the government is really free to acquire, deploy, expand existing surveillance technology outside of public view, without public input and without oversight. Often, as a result, we really don't know what's out there" (Stein et al., 2021).

What communities cannot see, they cannot resist. What they cannot resist, they begin to internalize as permanent.

When Helping Someone Gets You Watched

Surveillance architecture does not just monitor individuals. It maps associations. Brian Hofer, chair of the Oakland Privacy Advisory Commission, explained the mechanism directly: a single data point carries low privacy impact. But connect four location data points and you can identify 95% of people, according to a 2013 peer-reviewed study in *Scientific Reports* (as cited in Stein et al., 2021). Palantir's Gotham software—deployed in New Orleans starting in 2012—was designed precisely to find connections, trends, and social networks in data that would otherwise be invisible to human investigators (Stein et al., 2021).

The implication for community life is concrete. Helping someone—driving a neighbor to a medical appointment, attending the same mutual aid distribution, appearing repeatedly in the same location data as someone already flagged in a database—links you to their profile. In Chicago, the Strategic Subject List assigned risk scores partly based on the arrest records of people you knew, not crimes you committed (see Chapter 5). Association itself became evidence of risk.

This is the criminalization of care. Communities with the least access to institutional resources—hospitals, legal services, social workers—historically substitute those services with informal networks: neighbors who drive, friends who lend money, churches that feed people. Surveillance infrastructure, designed to detect social connections as threat indicators, treats these networks as suspicious by design. The mutual aid that allows communities to survive becomes data that marks people for enhanced monitoring.

Young-Drake et al. (n.d.) document a related dynamic: aggressive policing makes people less likely to call police when they need help. Crimes go unsolved. Violence escalates. The architecture that claims to produce safety produces the opposite—communities that stop calling for help because experience teaches them that police contact generates new risks.

Trust Does Not Survive a Surveillance Apparatus Designed to Map It

Former New Orleans Mayor Mitch Landrieu announced his administration's surveillance-backed gang intervention strategy in 2012 with explicit language: "We know who you are. We know who your friends are. We know where you go. We know everything about you" (Stein et al., 2021). The statement was directed at individuals the Palantir algorithm had flagged. But the message reached everyone. The surveillance state told communities that friendship, presence, and association were being documented and would be used.

Bruce Hamilton, former senior staff attorney for the ACLU of Louisiana, assessed his city's transparency record and reached a specific conclusion: "They don't deserve our trust" (Stein et al., 2021). That verdict came after New Orleans spent two years publicly denying it used facial recognition—while NOPD detectives were, in fact, running facial recognition searches through a state fusion center. The city fought a public records lawsuit to the Louisiana Supreme Court rather than release a map of its own publicly visible cameras (Stein et al., 2021).

Trust between communities and governing institutions does not erode abstractly. It erodes through specific experiences of deception, of surveillance deployed without consent, of data collected without disclosure. When residents discovered the camera that appeared above their community garden was connected to a Real Time Crime Center feeding algorithms they were never told about, the violation was not only technological. It was relational.

Surveillance Spending as Economic Extraction

The money question rarely appears in surveillance debates. It should. When a city deploys acoustic sensors in a Black neighborhood, the transaction is not neutral—it is a fiscal choice that redirects public resources toward surveillance infrastructure and away from everything communities have documented they need.

ShotSpotter costs between $65,000 and $90,000 per square mile annually (VerBruggen, 2025). Detroit—a city more than 80% Black, with surveillance deployment concentrated in its Black neighborhoods—maintains a ShotSpotter contract running approximately $2 million per year (VerBruggen,

2025). That contract represents less than 1% of the department's $400 million budget, a figure the technology's defenders cite as evidence the cost is negligible. But the question is not whether $2 million is small relative to total police spending. The question is what $2 million would purchase if directed elsewhere in a city where residents have documented decades of disinvestment.

Newburgh, New York City Councilperson Omari Shakur asked it plainly after his city renewed its contract: "You know how many people we can hire in our community with that?" (Mascia, 2024). His argument was not abstract. Eric Piza, a criminology professor at Northeastern University who has published peer-reviewed studies on ShotSpotter, stated directly that "hiring more police officers to conduct community policing and strategies like problem-oriented policing frankly have much better track records for preventing crime than technological solutions do" (Mascia, 2024). The money spent on sensors is money not spent on those alternatives.

The fiscal extraction extends beyond direct contract costs. A 2022 *NBC News* investigation documented that SoundThinking, the company behind ShotSpotter, actively lobbied Congress and federal agencies for grants while simultaneously guiding local police departments through the process of obtaining that funding (Mascia, 2024). The company positioned itself at both ends of the money pipeline—shaping federal grant availability and then helping cities access it to pay for the company's own product. Phillipsburg, New Jersey—population 15,000, five total shootings in a decade—deployed $297,000 in federal COVID relief funds on ShotSpotter sensors (Mascia, 2024). That is pandemic recovery money, intended to address the documented economic damage COVID inflicted on low-income communities, converted instead into surveillance infrastructure.

A *Trace* analysis identified 62 cities using ShotSpotter that averaged fewer than one shooting per month in which someone was injured or killed over a ten-year period (Mascia, 2024). Forty-three of those cities were in states that received top grades from gun violence prevention organizations for their existing restrictions. The technology was not deployed in response to documented need. It was deployed because the money was available and the company's sales infrastructure made it accessible.

The False Alert Tax

Surveillance systems impose costs not only in dollars but in time, safety, and community exposure to armed police presence. New York City's 2024 comptroller audit found that police were dispatched to alerts not actually involving shootings 87% of the time (Mascia, 2024). Chicago's inspector general found that in 2020 and early 2021, only about 9% of ShotSpotter alerts led to evidence of a gun-related criminal offense—roughly 1 in every 11 dispatches (VerBruggen, 2025). Each of those non-productive dispatches sends armed officers into a neighborhood with heightened tactical expectation.

The MacArthur Justice Center documented the consequence directly in a 2022 class action lawsuit against Chicago: "Every one of these deployments creates a dangerous, high-intensity situation where police are primed by ShotSpotter to expect to find a person who is armed and has just fired a weapon.

Residents who happen to be in the vicinity of a false alert will be regarded as presumptive threats, likely to be targeted by police for investigatory stops, foot pursuits, or worse" (as cited in Mascia, 2024). The center cited the case of Michael Williams, who spent nearly a year in jail on a murder charge based on a misinterpreted ShotSpotter alert.

The false alert infrastructure provides the mechanism through which police contact is generated—systematically, automatically, at scale. Chapter 7 documented what repeated police contact does to the body's stress response. The two are not separate phenomena. The acoustic sensor and the elevated cortisol reading are part of the same architecture.

Concentration, Not Coincidence

A 2024 *Wired* investigation using leaked sensor location data from more than 25,000 ShotSpotter sensors found that nearly 70% of people living in census block groups with at least one sensor identified as either Black or Latino, and nearly three-quarters of those neighborhoods were majority nonwhite (VerBruggen, 2025, citing Mehrotra & Scott, 2024). SoundThinking's response is that sensors are placed based on objective historical crime data. The problem with that defense is precisely the argument this book has made across eight chapters: historical crime data reflects historical policing patterns, which reflect historical decisions to concentrate enforcement in Black and Brown neighborhoods. Using that data to determine where surveillance goes ensures that surveillance follows the same logic as the discrimination that produced it.

Somerville, Massachusetts, Councilor Willie Burnley Jr. named the community cost: "Just because ShotSpotter is free does not mean that we should place microphones all across our community, particularly in communities of color, to surveil people" (Mascia, 2024). Somerville gets its ShotSpotter through a Department of Homeland Security grant at no direct cost to the city. The sensors are still disproportionately concentrated in communities of color. The absence of a direct budget line does not mean the absence of cost—it means the cost has been externalized onto the communities bearing the surveillance.

Baltimore: When the Sky Becomes Infrastructure

Baltimore's aerial surveillance program—planes carrying cameras that captured 32 square miles in a single image, recording everyone's movements for hours at a time—represented the logical end point of the surveillance expansion documented throughout this book. The system was operated by Persistent Surveillance Systems and funded in part through private donations. Baltimore residents were not informed it was operating until journalists reported its existence (see Chapter 2).

The Fourth Circuit Court of Appeals ruled it unconstitutional in *Leaders of a Beautiful Struggle v. Baltimore Police Department* (2021), finding that tracking an entire city from aircraft violated reasonable expectations of privacy. The ruling came after the surveillance had already operated. People had been tracked without consent before any court decided that tracking them was illegal. The pattern that recurs across this book—deploy first, litigate later, expand elsewhere while courts process previous violations—played out in Baltimore in its most literal form: the city itself became the surveillance apparatus.

What Baltimore's aerial program did to community behavior during its operation is not fully documented in available research. That research gap is itself significant. The absence of data on how communities change under surveillance systems rarely reflects the absence of harm—it reflects the absence of investment in measuring it.

The Fabric Holds—But Only in Resistance

Surveillance does not automatically dissolve community. What it does is force communities to spend resources surviving it that would otherwise go elsewhere. The organizing, the public records lawsuits, the coalition meetings, the City Council testimony—Eye on Surveillance in New Orleans, Oakland Privacy in California, Leaders of a Beautiful Struggle in Baltimore—all of that labor is a surveillance tax paid in time and capacity (Stein et al., 2021).

Nathan Sheard, Associate Director of Community Organizing for the Electronic Frontier Foundation, identified the structural dynamic: "Some of those traditional checks and balances as far as budgeting are being circumvented and police departments are building huge arsenals of surveillance technology without any kind of engagement from lawmakers and without the awareness of the public" (Stein et al., 2021). The public cannot regulate what it does not know exists. Communities cannot protect social space that has already been converted into data infrastructure without their knowledge or consent.

The social fabric is not simply torn by surveillance. It is redirected—toward surviving the watching, and away from everything else that communities need to build.

Chapter 9: Caught Before the Bell: How School Surveillance Turns Children Into Suspects

On January 29, 2019, sixteen-year-old Dnigma Howard was removed from class at Marshall Metropolitan High School in Chicago—not because she posed a threat to anyone, but because she had a cell phone. Her teacher called two school resource officers. What happened next was captured on the school's own security cameras: the officers tackled Dnigma down a flight of stairs, beat her, and tased her in front of classmates and teachers who could do nothing but watch. Before that video became public, the Cook County State's Attorney's Office had charged Dnigma with two counts of felony aggravated battery—a charge that could have meant expulsion and jail time for a teenager whose only infraction was a phone (Froelich, 2021).

Dnigma Howard's case did not happen in isolation. It happened inside a school system that had already decided certain children were problems to be managed rather than people to be educated. Surveillance technology did not produce that decision—but it amplifies it, memorializes it, and builds its logic into automated systems that operate long after any individual officer, administrator, or camera operator has moved on. This chapter examines how surveillance infrastructure inside schools compounds existing racial disparities in discipline, how it shapes the psychological development of children who never know unwatched space, and how the data generated in classrooms follows young people into adulthood as a permanent ledger of suspicion.

Surveillance as School Architecture: Who Gets Watched and How

The deployment of surveillance technology in American schools did not begin with facial recognition. Metal detectors, security cameras, and school resource officers have saturated schools serving predominantly Black and Brown students since at least the 1990s, a period when zero-tolerance discipline policies spread rapidly following the 1994 Gun-Free Schools Act. That legislation required year-long suspensions for students who brought weapons to school. Almost immediately, districts expanded its logic to cover minor behavioral infractions—and the consequences fell disproportionately on students of color and students with disabilities. Suspensions and expulsions climbed. Arrests followed (Froelich, 2021).

The technology layers that followed did not disrupt this pattern—they digitized it. Rekor Systems' OnGuard platform, marketed to school districts as a safety solution, packages vehicle recognition software with tracking radar, license plate readers, and 24/7 social media monitoring of students. Its algorithms claim to detect "suspicious movement patterns," though the company provides no public explanation of what that classification means or how it is calibrated (Madhukar, 2019). Facial recognition systems are now used in schools to regulate building access and support law enforcement activity. Behavioral threat assessment programs—now mandated by law in at least nine states—use device monitoring, social media surveillance, and anonymous reporting systems to identify students deemed likely to pose future risks. Disability justice advocates have documented how these assessments perpetuate racial profiling and increase student contact with police (Okoh, 2024).

The students who bear the weight of this infrastructure are not randomly distributed. Black students constitute 16% of the national school population but account for 31% of school-based arrests. Students with disabilities comprise 12% of school enrollment and represent 25% of school-based arrests and referrals to law enforcement. Schools with school resource officers show measurably higher arrest rates for discretionary offenses—disorderly conduct, trespassing, minor altercations that in other settings would not generate a criminal record (Froelich, 2021). Black girls face out-of-school suspension at six times the rate of their white peers (Madhukar, 2019). The surveillance does not create these disparities from nothing. It accelerates a system already pointed in one direction.

The Compounding Effect: When Every Space Is a Watched Space

A child growing up in a heavily surveilled urban neighborhood does not encounter surveillance only in school. She walks past cameras on the block, passes through metal detectors at the school entrance, sits under ceiling cameras in the classroom, carries a monitored school-issued device home, and is potentially subject to behavioral threat assessments that collect her social media posts. Her neighborhood may be subject to predictive policing algorithms. ShotSpotter acoustic sensors may monitor sound outside her bedroom window. She has never, in any meaningful sense, experienced unobserved space (Okoh, 2024).

Sociologist Ruha Benjamin has described this configuration as the "New Jim Code"—the use of data-driven systems to engineer new infrastructure for structural racism and social inequality, cloaked in the apparent objectivity of algorithmic decision-making (as cited in Okoh, 2024). In schools, that infrastructure is embedded in the physical environment. The cameras are built into the walls. The monitoring software runs in the background of every assigned device. The data is collected before any incident occurs.

Privacy International's (2024) investigation into education technology found that EdTech systems are increasingly oriented toward monitoring, security, and control rather than instruction. The report documented how "continual policing" makes it harder for students to develop the trust-based relationships that education requires, and that surveillance makes students reluctant to seek help from teachers they might otherwise confide in—not because they distrust the teacher, but because they understand their digital footprint will be judged instead of their actual experience. One student described feeling "constantly in a state of alert, afraid to make even the smallest mistake" (as cited in Madhukar, 2019). This is not a description of safety. It is a description of threat.

Research on the psychological effects of security technology documents a consistent pattern. Constant surveillance generates stress, hypervigilance, and what Malik et al. (2024) describe as "surveillance-related stress and paranoia:" conditions that erode cognitive function, impair emotional regulation, and compromise the conditions necessary for learning. These findings hold for adults subjected to surveillance environments; their implications for children, whose neurological and psychological development is ongoing, are more acute and more poorly studied. What is documented is behavioral: students who experience pervasive surveillance modify their conduct, self-censor, disengage from classroom participation, and avoid drawing attention to themselves in ways that reduce their educational engagement (Privacy International, 2024).

Lupton and Williamson (2017) traced the datafication of children from before birth through the school years, documenting how monitoring practices generate detailed dossiers on young people who have not consented—and in many cases whose parents do not fully understand what is being collected or how it is used. They found little evidence that any instruments specifically designed to protect children's rights in relation to dataveillance had been developed or implemented. That observation, made in 2017, remains largely accurate in 2024.

Permanent Records: How School Surveillance Writes the Future

The data generated by school surveillance systems does not stay in school. Behavioral risk assessments produce records that can be shared with law enforcement. Disciplinary histories follow students into juvenile justice systems and into adulthood. School-based arrest records appear on background checks. In some jurisdictions, students flagged by predictive analytics tools have been added to gang databases—databases that are almost impossible to exit, that affect housing and employment eligibility, and that can be shared across agencies without the subject's knowledge (Okoh, 2024).

The ACLU has characterized the school-to-prison pipeline as a disturbing national trend wherein children are funneled out of public schools and into the juvenile and criminal justice systems—a process driven by discretionary discipline policies applied unequally across racial lines, zero-tolerance rules, and the wholesale criminalization of school conduct (as cited in Froelich, 2021). Surveillance technology does not slow this pipeline. It adds automation to it. When a behavioral threat assessment algorithm flags a student before any incident occurs, when a social media monitoring tool reports a post to law enforcement, when an anonymous tip system generates a file that follows a child for years—the pipeline does not just process students after they act. It identifies and tracks them in advance.

Children who have been arrested at school face measurably worse long-term outcomes: lower graduation rates, higher rates of adult incarceration, diminished employment prospects. Each of these consequences is itself a data point that the next algorithmic system will treat as evidence of risk. The loop is closed. A child who was surveilled, flagged, disciplined, arrested, and charged becomes an adult whose record confirms, to the next system that encounters him, that surveillance was justified.

This is what stolen futures looks like in practice. Not a single dramatic moment of injustice—though those happen too, as Dnigma Howard's case makes plain—but an accumulation of records, assessments, flags, and disciplinary actions that compound across years until the path out becomes narrow or invisible. The surveillance did not cause the poverty or the underfunded schools or the discriminatory discipline policies. But it industrialized them, made them harder to contest, and embedded them in systems that claim the authority of objectivity.

Normalization and What It Costs

Children who grow up under pervasive surveillance normalize it. They learn, because they must, to treat being watched as the default condition of existence. They do not know an alternative. What the research has not yet fully reckoned with—because it cannot easily be measured—is what this normalization costs in terms of political imagination. Children who have always been watched may grow into adults who accept surveillance as natural, who lack the experiential basis for understanding what unwatched public space feels like, who are less equipped to contest monitoring systems that claim to be protective.

Privacy International (2024) identified this as the most pernicious dimension of surveillance-oriented education technology: it substitutes safety for order and learning for monitoring, with an opacity that makes the substitution hard to identify or challenge. When the camera is in the wall and the monitoring software runs invisibly, when the behavioral threat assessment is conducted by professionals who sincerely believe they are helping, the coercion is structural rather than interpersonal. There is no individual to hold accountable. There is only the system.

What is owed to the children processed by these systems is not primarily a policy fix—though policy matters, and Chapter 12 addresses the legislative terrain in detail. It is recognition that they were not served by the infrastructure nominally built to protect them. Dnigma Howard was not protected. She was tackled down a flight of stairs and charged with a felony for having a phone. The cameras in that school recorded everything. The data exists. And the question the data cannot answer—what it cost a sixteen-year-old girl to be hurt in front of her classmates, to be charged with crimes she did not commit, to carry that in her body—is precisely the question the surveillance system was never designed to ask.

Part III: Resistance and Refusal

Chapter 10: Reclaiming the Block: How Communities Fought Back Against Algorithmic Redlining

Chapter 10: Refusing the Lens: How Communities Fight Back Against Algorithmic Policing

On June 23, 2020, Justin Cummings stood before the Santa Cruz City Council and voted to ban the very technology his city had helped create. Cummings was Santa Cruz's first Black mayor. The city was the birthplace of PredPol, the predictive policing company whose algorithm had spread to departments across the country after a 2011 pilot in Santa Cruz itself. Two county supervisors had helped raise venture funding for PredPol. One had served on its board. When Cummings and his colleagues voted unanimously to prohibit the use of predictive policing and facial recognition by the city's police department, he was direct about why. "Understanding how predictive policing and facial recognition can be disproportionately biased against people of colour," he said, "we officially banned the use of these technologies in the city of Santa Cruz" (Santa Cruz Works, 2020). No speaker opposed the ordinance. Santa Cruz became the first city in the United States to ban predictive policing—and it did so by turning against a technology it had incubated (Governing, 2021).

That reversal is the subject of this chapter. Communities across the country have built an architecture of resistance against algorithmic surveillance that includes legal challenges, legislative campaigns, community organizing, technical countermeasures, and cultural refusal. This chapter maps that architecture, documents what has worked, names where limits remain, and examines what the pattern of resistance reveals about the nature of the system being resisted.

From Courtrooms to City Councils: Legal and Legislative Victories

The most concrete legal win against facial recognition produced not a ban but a set of structural constraints enforceable in federal court for four years. In April 2021, the ACLU, the ACLU of Michigan, and the University of Michigan Law School's Civil Rights Litigation Initiative filed a federal lawsuit on behalf of Robert Williams—wrongfully arrested at his Farmington Hills home in January 2020 after Detroit police ran a grainy surveillance image through a facial recognition system and acted on a match that was wrong (ACLU, 2024a). The complaint named the detective who misled a magistrate judge in his warrant application, the City of Detroit, and the Detroit Police Chief. Three years of discovery exposed what the ACLU described as systemic and comprehensive failures in how the department had been using the technology—failures that extended to Michael Oliver, wrongfully arrested in 2019, and Porcha Woodruff, eight months pregnant, wrongfully arrested in 2022 (Metro Times, 2024). All three were Black.

The June 2024 settlement prohibited Detroit police from arresting anyone based solely on a facial recognition match or on a photo lineup conducted immediately after a facial recognition search. It mandated officer training on the technology's elevated misidentification rates for people of color. It ordered an audit of every warrant obtained through facial recognition since 2017. The court retained jurisdiction to enforce the agreement for four years (ACLU, 2024a). What Williams did not get was what he wanted most: a ban. The department still uses the technology. The settlement is regulation—the strongest in the country—but it is not abolition.

Research on algorithmic accountability helps explain why evidence access was decisive in this case. Stewart's (2025) analysis of 168 legal cases involving algorithmic decision-making found that cases with evidence access succeeded at rates approaching 97% in domains without liability shields, while those without such access almost never succeeded—approximately 9%. For Williams, compelling discovery of the department's operational history converted one documented wrongful arrest into proof of a systemic pattern. Explanation of the algorithm's outputs was insufficient; access to what the department had actually done with those outputs over years was what made contestation possible (Stewart, 2025).

Legislative campaigns have produced parallel results, though with significant variation in durability. On May 14, 2019, San Francisco's Board of Supervisors voted 8–1 to pass the Stop Secret Surveillance Ordinance—the first ban on facial recognition by a major American city (TechCrunch, 2019). The ordinance prohibited any city department, including the police, from obtaining, retaining, accessing, or using facial recognition technology. It also required departments to seek public approval before acquiring any new surveillance equipment. Supervisor Aaron Peskin, who introduced the ordinance, framed it as an accountability measure: "We have a fundamental duty to safeguard the public from potential abuses" (WEF, 2019). The ban's supporters included the ACLU, the Electronic Frontier Foundation, and local organizers who had spent years documenting the technology's racial impact. It inspired more than 20 other

U.S. cities and counties to enact similar laws (SF Standard, 2024).

Those cities include Boston, Minneapolis, and Portland, Maine. By early 2025, 15 states had enacted some form of facial recognition regulation (Stateline, 2025). But the San Francisco case also illustrates the durable problem with city-level bans: five years after the ordinance passed, the SFPD had violated it at least twelve times by asking neighboring agencies to run searches on its behalf, and had omitted those violations from required annual transparency reports. Secure Justice, whose director Brian Hofer helped draft the original ordinance, sued the city for enforcement (SF Standard, 2024). Bans create accountability surfaces. They do not automatically change institutional behavior.

When Constitutional Litigation Becomes the Barrier

The xAI lawsuit against Colorado's Senate Bill 24-205, filed April 9, 2026, in the United States District Court for the District of Colorado, represents a harder version of the problem San Francisco's experience illustrates. SFPD violated the city's facial recognition ban procedurally — routing searches through neighboring agencies to avoid direct compliance. That is circumvention after the fact. The xAI complaint attempts something structurally different: establishing a constitutional floor below which anti-discrimination accountability requirements cannot reach at all, before they take effect. Count I of the complaint argues that Colorado's requirement that AI developers measure and disclose algorithmic discrimination in employment, housing, and healthcare decisions violates the First Amendment and "cannot survive any level of heightened scrutiny" (X.AI LLC v. Weiser, 2026, Compl. para. 128). If the Tenth Circuit accepts that argument, every CCOPS ordinance, every transparency requirement, every disclosure mandate that this chapter has documented as a resistance success faces the same challenge on appeal.

The structural limit of CCOPS ordinances identified earlier in this chapter — that they regulate acquisition but cannot compel decommissioning of systems already in place — now has a companion. Constitutional litigation can prevent accountability mechanisms from being enacted at all, not merely circumvented after the fact. The complaint also discloses that the White House established an "AI Litigation Task Force whose sole responsibility shall be to challenge State AI laws inconsistent with the policy" framework supporting AI industry interests (X.AI LLC v. Weiser, 2026, Compl. para. 52). That is a standing federal litigation infrastructure dedicated to dismantling the legislative architecture communities have spent years building. Organizers who drafted CCOPS ordinances, lawyers who won the Williams settlement, city councils that voted unanimously to ban predictive policing — all of them now face a coordinated federal legal apparatus funded specifically to undo those wins. The resistance documented in this chapter is real and consequential. It is also operating against an opposition that has moved to the constitutional level.

Force Them to Ask First: The CCOPS Model and Its Limits

Behind the ban campaigns sits a more durable oversight infrastructure. Community Control Over Police Surveillance (CCOPS) ordinances require government agencies to seek public approval before acquiring surveillance technology, disclose what tools they already possess, and develop use policies subject to community review. Oakland's Privacy Advisory Commission—which recommended the city's unanimous December 2020 ban on predictive analytics and biometric surveillance—operated under this framework. Commission chair Brian Hofer articulated the structural problem these ordinances are designed to interrupt: "Based on past data which was likely collected in a disproportionate manner as certain communities are policed more than others and certain crimes enforced more than others, a computer algorithm directs police back to the very same community being over-policed because 'that's where the crime is occurring'" (as cited in Governing, 2021). CCOPS ordinances force that argument into the public record before technology is purchased—not after it has produced wrongful arrests.

The model has spread significantly. Santa Clara County passed a predecessor ordinance in 2016. San Francisco's 2019 ordinance extended it. The core mechanism—surveillance impact report, public hearing, legislative approval—has been adopted in jurisdictions from Nashville to Palo Alto. Oakland took it further, becoming the first city to incorporate both predictive policing and biometric surveillance prohibitions into a comprehensive CCOPS framework, ensuring police cannot acquire any new surveillance technology without council permission following public input (EFF, 2021). Its power lies in changing the default: rather than surveillance tools entering departments through procurement processes insulated from public scrutiny, CCOPS makes acquisition a political question communities can contest before the technology is operational. The structural limit is also procedural: CCOPS ordinances regulate acquisition, not inherited infrastructure. They catch new purchases. They do not compel decommissioning of systems already in place.

Mutual Surveillance: Communities Watching the Watchers

Communities have not waited for legislation. Across the country, organizers have developed systematic counter-surveillance practices—documenting police technology deployment, mapping camera infrastructure, building public evidence of how surveillance systems operate in specific neighborhoods. The Ban the Scan campaign has organized residents to photograph and map facial recognition cameras, submit public records requests for procurement contracts, and build coalitions between civil liberties organizations, faith communities, and neighborhood groups (BanTheScan.org, n.d.). In Chicago, activists documented ShotSpotter sensor locations on the South and West Sides—the same predominantly Black and Latinx neighborhoods identified in Chapter 2—and used that documentation to challenge the technology's accuracy before the city council.

Countermapping practices adapt methods from environmental justice organizing, where communities have long had to make visible the infrastructure imposed on them in order to contest it. Applied to surveillance, counter-mapping produces spatial evidence that official narratives erase: the concentration of cameras in specific neighborhoods, the overlap between historical redlining maps and current camera density, the locations of automated license plate readers in relation to Black and Latino residential areas. This documentation does not by itself dismantle infrastructure, but it provides the evidentiary foundation for both litigation and legislative campaigns—and it shifts the epistemic burden. Defenders of surveillance systems must explain documented patterns of deployment rather than simply asserting technical neutrality.

You Cannot Encrypt a Body: Why Technical Fixes Fall Short

Technical countermeasures against facial recognition range from the widely documented to the largely theoretical. Artist and technologist Adam Harvey's CV Dazzle project demonstrated that certain patterns of high-contrast makeup and hairstyling could defeat early facial recognition algorithms by disrupting the spatial relationships between facial features those systems relied on (Harvey, 2010). The project attracted significant attention and generated related work in adversarial fashion—clothing and accessories designed to confuse machine learning systems. Research demonstrated that printed adversarial patches could cause image classifiers to produce confident misidentifications of physical objects (Brown et al., 2017).

The limits are real. As systems have grown more sophisticated, CV Dazzle-style approaches have become less reliable against state-of-the-art recognition. More fundamentally, individual countermeasures address only one node of a multilayer surveillance infrastructure. Defeating a facial recognition camera does not stop license plate readers, ShotSpotter sensors, or the data integration platforms that aggregate all these streams. Encryption protects communications; it cannot encrypt a body in public space. The ACLU has been direct on this point: the strongest protection is not obfuscation but abolition of the infrastructure itself (ACLU, 2024b). Technical resistance is most useful as documentation of surveillance's limitations—and as political argument that the technology is neither reliable nor inevitable—rather than as dependable individual protection.

Cultural Resistance: Refusing to Be Driven From Public Life

There is a form of resistance that operates below the level of law and technology: the refusal to allow surveillance to determine who occupies public space. When Beyoncé opened the Super Bowl 50 halftime show in February 2016 with a formation of dancers dressed as Black Panthers, performing before a nation that had spent the preceding year debating the legitimacy of Black Lives Matter, the performance was understood immediately as a claim on public visibility. The song "Formation" and its video, released the previous day, depicted images of Black New Orleans after Katrina—a surveillance and abandonment narrative—alongside defiant assertions of joy, solidarity, and Black cultural life that refused reduction to a threat to be managed (Caramanica et al., 2016). This is not surveillance resistance in the technical sense. It is the insistence that surveillance does not get to define the terms of Black existence in public.

Community organizing campaigns have incorporated this logic directly. Activists have organized dance parties and celebrations in front of surveillance cameras in heavily monitored neighborhoods, documenting the absurdity of systems designed to track criminal behavior in the presence of collective joy. Afrofuturist art movements have produced visual and narrative work that imagines Black futures outside surveillance infrastructure—not as escapism but as political imagination, asserting that the present configuration of watched and watchers is not fixed. Hip-hop's documentary tradition—from N.W.A's "F*** tha Police" in 1988 to J.

Cole's "m y . l i f e" in 2021—has created a sustained counter-record of what surveillance and aggressive policing do to specific communities, outside the terms of official data systems.

What unifies these cultural practices with legal campaigns and organizing is a single underlying claim: that communities subjected to surveillance have the standing, the knowledge, and the right to determine what monitoring is imposed on them. Brian Hofer made that claim before an Oakland city council. Organizers make it every time they document a camera, file a public records request, or assemble in a watched public square and refuse to leave. The resistance is not one campaign. It is an insistence, accumulated across fronts and years, that the surveillance architecture is a political choice—and another choice is available.

Chapter 11: Ungovernable by Design: Abolitionist Technology Frameworks and the Future of Public Safety

Beyond Reform—Reimagining Public Safety Without Surveillance

On October 11, 2022, after postponing the vote six times under sustained community pressure, Detroit's city council voted 5–4 to approve a $7 million expansion of ShotSpotter's acoustic surveillance network across ten additional neighborhoods. The debate had run for months. Hundreds of Detroiters packed council chambers week after week. A coalition including the ACLU of Michigan and the Detroit Justice Center held a press conference the morning of the vote demanding rejection. Community organizer Alyx Goodwin of the Action Center on Race and the Economy had laid out the alternative argument directly: "If people had jobs, money, after-school programs, housing, the things that they need, that's going to reduce gun violence" (The Intercept, 2022).

Council approved it anyway. Council member Gabriela Santiago-Romero, who voted no, identified the tradeoff precisely: "If council is going to vote yes on ShotSpotter, we should be voting yes on community-based intervention programs and fully funding them. We should vote yes on truly affordable housing. We should vote yes on mental health resources" (Michigan Public Radio, 2022). Branden Snyder of Detroit Action called the vote "a slap in the face to thousands of Detroit residents teetering on the precipice of homelessness, economic instability, and joblessness" (Detroit News, 2022).

The question of what public money should do for Detroit's communities—whether it should build surveillance infrastructure or address the deprivations surveillance claims to manage—is the central tension this chapter examines. But Detroit is no longer the hardest case. That distinction now belongs to what happened in Colorado.

When Constitutional Litigation Becomes the Barrier

Communities across the country have built legislative wins against surveillance. San Francisco banned facial recognition in 2019. Oakland followed with predictive policing and biometric surveillance prohibitions. Santa Cruz—the technology's birthplace—voted unanimously to prohibit the very systems it had incubated. These victories took years of organizing, legal strategy, and political pressure. They were not inevitable.

On April 9, 2026, before Colorado's anti-discrimination requirement for AI systems could take effect, xAI LLC filed a federal lawsuit in the United States District Court for the District of Colorado arguing that the law violates the Constitution. The complaint's 75 pages and six counts of relief illustrate a problem that CCOPS ordinances and facial recognition bans did not fully anticipate: constitutional litigation as a preemptive veto.

The difference between San Francisco's experience and what the xAI complaint represents is structural. SFPD violated the city's facial recognition ban by routing searches through neighboring agencies—circumventing the rule after it was enacted. The xAI complaint attempts to establish a constitutional floor below which anti-discrimination accountability requirements cannot reach at all, *before they take effect*. Count I argues that Colorado's requirement that AI developers measure, disclose, and mitigate algorithmic discrimination in employment, housing, and healthcare decisions constitutes compelled speech that "cannot survive any level of heightened scrutiny" (X.AI LLC v. Weiser, 2026, Compl. para. 128).

The complaint's First Amendment argument is straightforward in its framing: Colorado is forcing xAI to adopt "ideological views" about fairness and equity that the company does not share (X.AI LLC v. Weiser, 2026, Compl. para. 3). Requiring developers to measure whether their systems produce disparate impacts based on race, the company argues, is equivalent to requiring them to speak messages they would not otherwise make. If accepted, that argument does not affect only Colorado's law. Every disclosure requirement enacted by the cities and counties documented in Chapter 10—every CCOPS ordinance, every transparency mandate, every accountability mechanism that emerged from sustained community organizing—faces the same constitutional challenge on appeal.

The complaint also discloses infrastructure that communities may not have fully understood they were opposing. The White House, in a December 2025 executive order, "established an 'AI Litigation Task Force whose sole responsibility shall be to challenge State AI laws inconsistent with the policy'" framework supporting AI industry innovation (X.AI LLC v. Weiser, 2026, Compl. para. 52). That is a standing federal litigation apparatus, funded specifically to dismantle the legislative architecture communities spent years building. Organizers who drafted CCOPS ordinances, lawyers who won the Robert Williams settlement in Detroit, city councils that voted unanimously to ban predictive policing—all now face coordinated federal legal machinery designed to undo those wins through constitutional doctrine rather than democratic opposition.

The structural limit of CCOPS ordinances identified in Chapter 10 has a federal companion. Constitutional litigation can prevent accountability mechanisms from being enacted at all, not merely circumvented after the fact. Resistance is real and consequential. It is also operating against an opposition that has moved to the constitutional level.

This matters because it establishes why regulatory approaches, even when communities successfully enact them, cannot solve the surveillance problem. The barriers are multiple. Some are political. Some are now constitutional. But the deepest barrier is theoretical: surveillance systems designed to manage populations already targeted by historical policing cannot be made "fair" through better disclosure or bias mitigation. Ruha Benjamin identified this problem years before the xAI lawsuit materialized.

Surveillance Cannot Be Optimized Into Equity

The first move in any conversation about surveillance reform is the claim that the technology just needs to be made better. Fix the bias in the training data. Require independent audits. Mandate human review of every match. Make the algorithm transparent. The previous chapters have documented why this frame is insufficient, and the theoretical foundation for why it fails was articulated clearly by Princeton sociologist Ruha Benjamin in *Race After Technology* (2019). Benjamin coined the "New Jim Code" to describe technologies that "reflect and reproduce existing inequities but that are promoted and perceived as more objective or progressive than the discriminatory systems of a previous era" (pp. 5–6).

Her taxonomy of discriminatory design identifies three patterns: systems that explicitly amplify hierarchies, systems that ignore and thereby replicate social divisions, and systems that aim to fix racial bias but ultimately do the opposite (Benjamin, 2019).

The third category is the most dangerous for reform-minded advocates. An unbiased facial recognition system is still a facial recognition system. A more accurate ShotSpotter still routes police to Black neighborhoods because those neighborhoods carry the historical arrest data that determines where the microphones are placed. A COMPAS risk assessment that perfectly predicts recidivism would still punish people for conditions—poverty, neighborhood, social networks shaped by prior over-policing—that the system did not create and cannot remedy.

Mariame Kaba, whose *We Do This 'Til We Free Us* (2021) became a *New York Times* bestseller and a primary text for organizers across the country, identified this pattern directly when discussing body cameras: paying for them, she argued, "is giving money into the very system you want to actually shrink" (NBC News, 2021). The problem is not that the cameras are inaccurate. The problem is that they are cameras in the hands of a system whose function is to surveil specific communities.

This distinction between equality and equity runs through the abolitionist critique of surveillance reform. Equality would mean everyone is surveilled the same. Equity would mean communities are not surveilled in ways that compound existing structural disadvantages. Neither version addresses the more fundamental question: whether surveillance of this kind is a legitimate instrument of public safety at all.

Kaba's argument is that "prisons, policing, and surveillance are part of a natural order that simply cannot be undone" only in the imagination of people who have not studied how recently these systems were constructed (as cited in Lateral Reviews, 2021, p. 1). They were built. They can be dismantled.

The xAI litigation illustrates why. The company frames anti-discrimination accountability as speech compulsion—a First Amendment violation. That framing cannot be defeated through better regulation. No law can be "fair" enough if the underlying premise is that measuring discrimination violates constitutional rights. The answer is not to optimize the law. It is to abandon the premise entirely: to recognize that surveillance systems designed around historical policing patterns cannot be reformed into neutrality because their purpose was never neutral.

The Cost of Watching: What Surveillance Spending Displaces

The investment-redirection argument is not an abstraction. It has a budget line. The Department of Homeland Security Urban Area Security Initiative provided $615 million to local and state agencies for counterterrorism activities in 2022 alone—funding that city police departments have used to purchase ShotSpotter contracts, social media monitoring software, drones, and license plate readers (MIT Technology Review, 2022).

Chicago entered a three-year, $33 million ShotSpotter contract in 2018, renewed it in 2021 despite a Chicago Inspector General finding that approximately 91 percent of ShotSpotter alerts produced no documented evidence of gun-related crime, and the city's own data showed the technology had not significantly reduced gun violence (Chicago Office of Inspector General, 2021). ShotSpotter disputes characterizing these outcomes as "false positives," arguing that an absence of documented evidence does not mean no gunshot occurred—but the operational consequence is the same: police dispatched tens of thousands of times to locations where no investigation was possible.

Atlanta discontinued its earlier ShotSpotter pilot after a 2019 internal cost-benefit analysis found the system yielded only five arrests and five gun recoveries across four beats over seven months — a cost ratio of $56,000 per arrest or gun recovered. A Miami-Dade Police Department memo concluded that "the cost outweighs the benefits of the ShotSpotter application," noting that "more than half of the time MDPD officers responded, they did not find anything: no scene, no evidence, and no gunman," a pattern consistent with Atlanta's own evaluation of its 2019 pilot.

These are not edge cases. They are the documented pattern of surveillance spending: systems that concentrate police activity in Black and Brown neighborhoods, generate alerts that put residents at risk, and consume public resources that could be redirected.

Branden Snyder named the contradiction precisely when Detroit debated using COVID relief money for ShotSpotter: "The Biden administration passed the American Rescue Plan and put forth this Covid relief money to inject money into local economies and get people back on their feet after the pandemic," he said. What it does instead, he argued, "is fatten the wallets of ShotSpotter" (The Intercept, 2022).

The surveillance economy's contracts run into the billions. Against that figure, community alternatives appear modest.

Community-Based Safety: The Evidence Base

CAHOOTS—the Crisis Assistance Helping Out On The Streets program run by White Bird Clinic in Eugene, Oregon—operated for 35 years at a fraction of police costs, diverting between 5 percent and 23 percent of emergency dispatch calls depending on methodology, and requiring police backup in only 1 percent to 2 percent of responses (White Bird Clinic, n.d.; National League of Cities, 2024; Eugene Police Department, 2019). CAHOOTS cost the city $820,586 under its 2022 contract, compared to the Eugene Police Department's $68 million budget—a ratio that illustrates how much community safety infrastructure can be built for what surveillance costs.

CAHOOTS is not hypothetical and not isolated. It spawned replications in Portland (Portland Street Response), Denver (Support Team Assisted Response, or STAR), and Olympia, Washington (Crisis Response Unit), among dozens of programs across the country (The Nation, 2025). By 2024, cities from Albuquerque to Anchorage were operating alternative response units, with hundreds more requesting information about the model (Marshall Project, 2024).

The most rigorous study of an alternative response program examined Denver's STAR program during its initial six-month pilot in 2020. Stanford researchers Thomas Dee and Jaymes Pyne analyzed crime data across all 36 Denver police precincts before and after STAR launched, using a quasi-experimental design. Their findings, published in *Science Advances* (2022), showed that STAR's operating districts saw a 34 percent reduction in low-level crime—approximately 1,400 fewer offenses—during the pilot period. Serious crime did not increase.

The cost of STAR response averaged $151 per incident, compared to $646 when individuals were funneled through the full criminal justice system (Dee & Pyne, 2022). By 2023, STAR responded to 7,000 calls for service and 46 percent of all eligible calls citywide, with a base budget of $7.2 million (National League of Cities, 2024).

The CAHOOTS model itself has generated significant data. University of Oregon researcher Nathan Burton found that CAHOOTS diverted approximately 23 percent of calls in the emergency response system from traditional law enforcement and that the program was preventing crises rather than only responding to them (University of Oregon, 2025). Two-person teams—a medic and a crisis worker—handled welfare checks, conflict resolution, substance use crises, and suicide intervention for 35 years without weapons, relying on "trauma-informed de-escalation and harm reduction techniques" (White Bird Clinic, n.d.).

In April 2025, Eugene ended its contract with CAHOOTS amid budget pressures and compliance disputes—a development researchers called alarming and advocates called a cautionary example of how community safety infrastructure loses funding battles against surveillance budgets (KVAL, 2025). The program continues operating in Springfield, Oregon.

Restorative justice programs have three decades of evidence showing reductions in recidivism, higher victim satisfaction, and lower costs than incarceration-based approaches. The Marshall Project's 2024 survey of alternative response programs across Albuquerque, Durham, Los Angeles, Dayton, and Anchorage found that unarmed responders required police backup at rates of 1 percent to 5 percent—and that no known major injuries to community responders had occurred.

The Reform Versus Abolition Tension—and Why Black Communities Are Not Monolithic

The abolition framework meets real resistance, including within Black communities most directly affected by surveillance. That resistance deserves honest treatment rather than erasure. Kaba acknowledged this directly: PIC abolition "is an unpopular view," she said in 2021, even as the George Floyd uprising brought abolitionist thinking to mainstream attention (NBC News, 2021).

Some Black elected officials, community leaders, and residents in heavily surveilled cities have supported the expansion of police technology on the grounds that gun violence is immediate and alternatives are unproven. In Chicago, community members in neighborhoods with the highest homicide rates have at various points called for more police presence, not less. The organizers calling to end ShotSpotter and the residents calling for faster police response to shootings are often operating in the same neighborhoods, drawing on different experiences of what the surveillance system has done to them.

Kaba's framework does not pretend this tension does not exist. Her argument is that communities have been given a false choice—between the violence of under-policing and the violence of over-policing—rather than the actual investments in housing, healthcare, schools, and economic stability that evidence shows reduce violence without surveillance. "When we talk about prison-industrial complex abolition," she said, "we're talking about a dual project… dismantling death-making institutions, like policing and prisons and surveillance, and creating life-affirming ones, putting resources and investing in the things we know do keep people safe—housing, healthcare, schooling" (Democracy Now!, 2021).

Generational differences in this debate are documented if not fully quantified. Younger Black activists in the post-Ferguson, post-Floyd organizing wave have disproportionately embraced abolitionist frameworks. Older Black residents carry experiences of 1990s crack epidemic violence and the failures of that era's policing to protect them. Gender shapes the debate differently still: Black women have been disproportionately both the targets of surveillance systems and the architects of abolitionist organizing.

Acknowledging these differences is not a concession that abolition is wrong. It is recognition that building toward it requires the same attention to specific community conditions that characterizes all resistance strategies documented in Chapter 10.

Technology for Liberation: When Infrastructure Should Be Built

Abolition of surveillance infrastructure does not mean abolition of all technology in public safety. The distinction worth drawing is between technology that monitors and classifies populations versus technology that delivers services to them.

CAHOOTS dispatches a medic and a crisis worker via a 911-integrated system. That is technology for delivery.

ShotSpotter installs microphones in Black neighborhoods to track acoustic signatures and route police to locations. That is technology for surveillance.

One is technology for care. The other is technology for control.

Benjamin's framework points toward what she calls "socially just imaginaries"—design processes that prioritize equity over efficiency and social good over market imperatives (2019, p. 183). In practice, this looks like:

Community broadband in digitally underserved neighborhoods rather than smart city sensor networks.

Telehealth infrastructure for communities with limited access to mental healthcare rather than predictive risk assessment tools.

Crisis response dispatch systems built around mental health workers rather than around police.

These are not theoretical replacements. They are operating programs—some of which have been defunded in favor of surveillance contracts.

The abolitionist technology framework is not anti-technology. It is pro-accountability about whose interests technology serves. When Kaba calls for "creating the conditions for dismantling prisons, police, and surveillance" (as cited in Herskind, 2021, p. 1), the operative word is *conditions*—the housing, healthcare, economic stability, and community infrastructure that make surveillance-based policing less necessary because the underlying deprivations that enable its justification are addressed.

That project is long-term and contested. But as the evidence from Denver, Eugene, Portland, and dozens of other cities now shows, it is not speculative.

Chapter 12: The World Is Watching: International Solidarity and Global Resistance Against Surveillance States

Learning from Worldwide Fights Against Surveillance

On December 21, 2017—four days before Christmas—Ed Bridges was walking through the Queen Street shopping area in Cardiff when South Wales Police deployed its AFR Locate system, facial recognition cameras scanning passersby at up to 50 faces per second. Bridges was not suspected of any crime. His face was captured and run against a database of persons of interest. He was not on it. The scan lasted seconds. He never knew it happened until afterward. He sued anyway. By August 2020, the Court of Appeal of England and Wales had ruled that the deployment was unlawful—a violation of Article 8 of the European Convention on Human Rights, a failure of data protection law, and a breach of the Public Sector Equality Duty under the Equality Act 2010 (*R [Bridges] v. Chief Constable of South Wales Police*, 2020). It was the first successful legal challenge to automated facial recognition anywhere in the world.

One case in Cardiff. One person who was not arrested, not charged, not even stopped. The precedent reached across continents. The strategies that produced it, and the global fights against surveillance that preceded and followed it, are what this chapter examines.

Hong Kong 2019: Tactics the World Learned

The 2019 Hong Kong protests against the extradition bill were the most sustained confrontation with state surveillance infrastructure that any democratic resistance movement had mounted at that point. Protesters faced cameras equipped with facial recognition, AI capable of tracking individuals across networks, and a law enforcement architecture that could move from identification to arrest with minimal friction (CBC News, 2019).

Their response was layered and deliberate. Masks and face coverings became standard equipment, not primarily against tear gas but against cameras. Hand-held laser pointers were used to disrupt image capture; when police classified them as offensive weapons, protesters organized mass "stargazing" demonstrations outside Hong Kong's Space Museum, hundreds of people training lasers at the sky in collective defiance (UAB Institute for Human Rights, 2025). In August 2019, protesters dismantled smart lampposts in Kowloon suspected of housing surveillance equipment (Participedia, 2019). Communication discipline matched physical tactics through Telegram channels, hand-signal systems, and the decentralized "be water" strategy that made the movement resistant to targeted arrests of leaders.

The limits were real. China's imposition of the National Security Law in 2020 effectively ended the protest movement, and the surveillance apparatus remains. What Hong Kong demonstrated was not that counter-surveillance tactics can defeat a determined authoritarian state—they cannot—but that individuals facing surveillance can significantly impede identification, and that forcing surveillance infrastructure into public view is itself a form of resistance. The lesson for U.S.

organizers is not the specific tactics but the underlying principle: name the infrastructure, make it legible, force its defenders to justify it publicly, and insist on the right to occupy public space without being catalogued.

The UK: What a Courtroom Win Requires

Ed Bridges's case succeeded where many surveillance challenges fail, and the reasons matter as much as the outcome. The High Court initially dismissed his challenge in September 2019. The Court of Appeal reversed this in August 2020, upholding three of five grounds: the deployment lacked the legal framework required by Article 8(2) of the European Convention on Human Rights; the data protection impact assessment was deficient; and police had not taken reasonable steps to determine whether the software discriminated on the basis of race or sex (*R [Bridges] v. Chief Constable of South Wales Police*, 2020).

The ruling's significance lay in what it required. Facial recognition could not be deployed without a legal framework that was accessible, foreseeable, and that constrained police discretion. The absence of such constraints—not the technology itself—made the deployment unlawful. South Wales Police did not appeal.

This is the durable tension of litigation-based resistance: wins create accountability requirements, but accountability requirements can be fulfilled. By 2023, UK police forces were expanding live facial recognition deployments in London, arguing that updated policies now met the legal standard Bridges had established. The accountability surface the litigation created became something police could satisfy rather than an obstacle they could not pass. The strongest interpretation of Bridges is not a ceiling on police power but a floor of procedural obligation.

China and the Uyghur Surveillance Genocide: The Endpoint

Understanding what surveillance becomes without constraint requires looking directly at what China has built in Xinjiang. The system documented by UN investigators, Pulitzer Prize-winning journalists, and Human Rights Watch represents the most comprehensive deployment of biometric surveillance against an ethnic and religious minority in the contemporary world.

The infrastructure includes cameras with facial recognition, mandatory biometric data collection encompassing facial imagery, iris scans, and DNA samples, neighborhood "convenience police stations," and broad-spectrum monitoring of personal devices and financial activity (Bulletin of the Atomic Scientists, 2022). The UN Office of the High Commissioner for Human Rights found in August 2022 that the surveillance architecture underpinned "serious human rights violations," and that the allegations "may constitute international crimes, in particular crimes against humanity" (OHCHR, 2022). The U.S. genocide determination was issued in January 2021 and reaffirmed the following month.

The technology was not developed in isolation. American and Western technology components—including semiconductor chips from U.S. manufacturers—were incorporated into surveillance systems deployed against Uyghurs, documented through government bidding records and supply chain analysis (Radio Free Asia, 2021). The Uyghur Forced Labor

Prevention Act and entity-list designations of Chinese surveillance companies represent partial responses.

Xinjiang represents the logical conclusion of a trajectory: what happens when the same logic of demographic targeting and pre-emptive criminalization documented in Detroit and Chicago is applied without civil rights law, federal oversight, community resistance, or press freedom. The technology is not different in kind from what U.S. police departments deploy. The political context is.

Palestine: Testing Ground for Global Export

The surveillance infrastructure deployed against Palestinians in the occupied West Bank is documented in granular detail because of testimony from Israeli soldiers and investigations by Amnesty International, Human Rights Watch, and Breaking the Silence.

Blue Wolf is a smartphone application used by Israeli soldiers to photograph Palestinians and match images against a database described as the IDF's "Facebook for Palestinians" (Washington Post, 2021). Units were incentivized to gather the most photos, with prizes awarded for highest-volume collection—what Amnesty International's lead AI researcher described as a gamification of mass biometric surveillance (Amnesty International, 2023).

Red Wolf, documented in Amnesty International's 2023 report *Automated Apartheid*, removes human decision-making from checkpoints altogether. It automatically scans every face crossing a checkpoint, adds unrecognized faces to the database without consent, and assigns color-coded risk profiles that determine passage. Palestinian residents describe the effect directly: "They can tell you that your name is not in the database, as simple as that, and then you're not allowed to pass through to your house" (Amnesty International, 2023).

The surveillance does not merely monitor movement. It determines whether movement is possible at all.

The connection to U.S. policing is not metaphorical. Surveillance technology developed and refined in occupied Palestinian territories has been marketed globally as "field tested," with a client base that includes U.S. police departments (IMEU, 2021). When organizations like the Movement for Black Lives and Palestinian solidarity movements assert a shared analysis of surveillance as population control, they are identifying a structural relationship, not drawing an analogy.

India: Deployment Without Framework

India presents the case of a democracy deploying facial recognition at mass scale without comprehensive legal framework, against a background of documented caste and religious discrimination in law enforcement.

Delhi Police used facial recognition to identify more than 1,900 people in connection with the northeast Delhi communal violence of February 2020, comprising at least 73 percent of all riot-related arrests (The Wire, 2025). A 2022 RTI response confirmed facial recognition evidence was presented in over 750 riot cases, without Delhi Police specifying the legal provisions under which such evidence was admissible. More than 80 percent of those cases have since resulted in acquittals or discharges.

Targeting patterns are documented by Amnesty International's Ban the Scan Hyderabad campaign: Muslims, Dalits, Adivasis, and transgender people bear disproportionate surveillance (Amnesty International India, n.d.; Al Jazeera, 2022). Vidushi Marda of Article 19 articulated the structural mechanism: "If you are a Dalit woman in India, the nature and extent to which you are under surveillance are far more than an upper caste Hindu man" (Santoshini, 2022). She also noted that "there are so many subjective human decisions that are made before the technology even comes into play" (Santoshini, 2022).

India's proposed nationwide Automated Facial Recognition System has been contested through litigation and advocacy without yet being stopped.

The Constitutional Ceiling: What Federal Preemption Threatens

The international case studies share a structural feature that has now materialized domestically: the entities being regulated do not recognize the jurisdiction of the regulators.

In April 2026, xAI filed *X.AI LLC v. Weiser* to permanently block Colorado's Senate Bill 24-205, a disclosure requirement for AI systems used in housing, employment, and healthcare. The complaint alleged six constitutional violations, including First Amendment compelled speech and Commerce Clause extraterritorial regulation. The Commerce Clause argument is particularly significant: xAI, incorporated in Nevada and headquartered in California with no Colorado offices, argues that a state cannot impose accountability requirements on AI systems developed entirely outside its borders. If courts accept this, every AI system marketed nationally would be regulable only by the state where the developer is headquartered—Delaware or California for every major AI company (X.AI LLC v. Weiser, 2026, Compl. para. 161).

The December 2025 White House executive order named Colorado's law specifically and established an AI Litigation Task Force charged solely with challenging state AI laws inconsistent with its policy framework (X.AI LLC v. Weiser, 2026, Compl. para. 52). Federal resources are now deployed to defeat state-level accountability requirements.

The wins documented across this book are real—Robert Williams's settlement, the Santa Cruz ban, Oakland's CCOPS framework. But the international record, and the constitutional litigation now emerging domestically, establishes what communities building durable resistance must account for. What the global cases show is that this contest is not unwinnable. Ed Bridges won. Hong Kong protesters made surveillance visible at enormous cost. Palestinian activists built global solidarity that named the surveillance as apartheid. India's digital rights organizations created litigation frameworks that constrained deployment. The question is not whether resistance is possible. It is what it requires on this particular terrain, against these particular opponents, right now.

Building Global Solidarity: Shared Analysis, Local Strategy

Technology companies operate globally. NEC supplies facial recognition to DHS and to police departments on multiple continents. Palantir sells data integration to U.S. agencies and operates across Europe, Israel, and beyond. When South Wales Police deployed facial recognition, they used commercial software developed in the same transnational market that supplies U.S. law enforcement. When Indian police use facial recognition to identify protest participants, they draw on algorithmic architectures trained on datasets that overlap with those used in the United States.

Resistance does not yet operate at the same scale. The cases that have produced the most durable accountability—the Bridges ruling, the Detroit settlement, the Santa Cruz ban—emerged from sustained, locally rooted organizing that was simultaneously informed by a global analysis of what surveillance infrastructure is for.

The appropriate lesson from global solidarity is not to import tactics but to share analysis. The same companies, the same algorithms, and the same logic of demographic targeting and pre-emptive criminalization operate across these contexts. Understanding that is what allows communities to name what they are resisting—not a local police department's bad policy but a global infrastructure of control—and to build resistance proportionate to what they face.

Part IV: The Future isn't Written

Chapter 13: Possible Futures: Dystopia, Reform, or Liberation?

Why Surveillance States Are Not Permanent

In May 2024, Houston Mayor John Whitmire announced that his city would end its partnership with ShotSpotter, the gunshot detection network installed across Houston's neighborhoods. He was direct about his reasons: the technology was, in his words, "a gimmick, feel-good program" (StateScoop, 2024). The contract — $3.5 million over five years, not yet expired — was something Whitmire had decided Houston no longer needed. Seven months later, Baton Rouge cut its own ShotSpotter contract under a $9 million departmental budget reduction, ending a deployment that had been in place since 2007 (StateScoop, 2025). Chicago, which had spent more than $50 million on ShotSpotter by September 2024, let its contract expire despite a 33-14 city council vote to extend it (The Appeal, 2024). In fall 2025, Oak Park and Evanston, Illinois terminated license plate reader networks; Eugene and Springfield, Oregon decommissioned 82 combined cameras (EFF, 2025).

These cancellations are not the story of surveillance ending. New contracts are being signed. The federal surveillance infrastructure documented throughout this book — Mobile Fortify, ImmigrationOS, the Automated Targeting System — expands independently of municipal procurement decisions. But the pattern of cities abandoning surveillance contracts after years of documented failure, often under budget pressure, establishes something the surveillance industry does not advertise: these systems are not permanent. Their vulnerabilities are structural.

Where the Infrastructure Cracks

The economic fragility is the most immediate vulnerability. Atlanta's 2019 internal memo found ShotSpotter cost $56,000 per gun recovered and declined renewal (City of Atlanta, 2020). Chicago's inspector general found approximately 91 percent of ShotSpotter alerts produced no documented evidence of gun-related crime (Columbia Law Review, 2025). Cities spending millions annually on systems that produce no measurable safety benefit will eventually face constituents and auditors demanding a different accounting. Budget crises accelerate that reckoning — Baton Rouge's cut was driven explicitly by a $9 million departmental reduction, not a change in principle.

The technical vulnerabilities compound the economic ones. The vendor lock-in documented in Chapter 4 creates brittleness: DHS designed its biometric architecture so specifically around NEC's algorithms that switching vendors would require rebuilding critical systems at an estimated cost of $100 million and re-extracting 1.2 billion facial images over six to twelve months (Driver & Jean-Baptiste, 2024). Infrastructure that expensive to replace also becomes a liability when the single-source vendor is acquired, raises prices, or faces regulatory action. Systems engineered to resist competitive procurement are systems whose fragility concentrates at one pressure point.

The energy demands of AI surveillance constitute the least-discussed vulnerability. U.S. data centers consumed 183 terawatt-hours of electricity in 2024 — more than 4 percent of total U.S. electricity consumption — with demand projected to grow 133 percent by 2030 (Pew Research Center, 2025). Cornell University researchers estimated that current

AI infrastructure growth will produce 24 to 44 million metric tons of CO_2 annually by 2030, the emissions equivalent of five to ten million additional cars on U.S. roadways (Cornell Chronicle, 2025). At a moment when climate change is straining power grids and producing residential electricity price increases across multiple states, the energy demands of mass surveillance will increasingly compete with basic public services for the same constrained resources. The surveillance apparatus is not insulated from the material conditions of the world it claims to make safer.

The political vulnerabilities are less predictable but perhaps most significant. In March 2024, San Francisco voters passed Proposition E by 54.09 percent, expanding police surveillance authority and weakening the oversight mechanisms attached to the city's 2019 facial recognition ban (San Francisco Department of Elections, 2024). The vote demonstrated that surveillance bans can be reversed through democratic legitimation, and that crime-focused political campaigns can reframe accountability requirements as obstruction. The same year, EFF documented SFPD violating the surveillance ordinance at least twelve times — while the Prop E campaign was being organized (EFF, 2024). Two trends operating simultaneously: expansion through the ballot and circumvention regardless of what democratic institutions decide.

Four Possible Futures

Four trajectories are currently in motion. They are not mutually exclusive.

The *accelerated police state* is the trajectory in which the constitutional architecture under construction by xAI's lawyers, the White House AI Litigation Task Force, and federal surveillance expansion succeeds. State accountability laws are preempted or defeated in court. CCOPS ordinances govern only municipal procurement while federal surveillance expands without local check. Mobile Fortify operates in every major city. Facial recognition becomes routine at protests, during traffic stops, at school entrances. The wrongful arrest rate persists because there is no legally enforceable corroboration standard. Accountability for algorithmic harm is constitutionally insulated. The pattern this book has documented continues, accelerated.

The *managed surveillance state* is the trajectory in which regulatory frameworks survive constitutional challenge but produce procedural compliance rather than dismantling. Police departments acquire documentation requirements but continue using facial recognition. Cities adopt CCOPS ordinances and conduct annual audits confirming surveillance continues under new governance. Bridges-style procedural floors become the ceiling of required accountability. ShotSpotter contracts end but are replaced by integrated platforms that bundle license plate readers, AI report-writing tools, and acoustic sensors under different product names — which is precisely what SoundThinking accomplished as Chicago ended ShotSpotter while simultaneously piloting CrimeTracer from the same vendor (The Appeal, 2024). The infrastructure persists; the branding changes.

Abolition and technology refusal is the trajectory in which economic contradictions, technical failures, and organized community pressure converge to dismantle surveillance infrastructure rather than regulate it. This requires not one but multiple simultaneous conditions: budget failures making surveillance contracts politically costly; legal victories imposing real operational constraints; alternative safety programs that demonstrably work; and community organizing that makes surveillance itself a contested political identity rather than a neutral public safety option. Each condition exists in partial form. Denver's STAR program, CAHOOTS, and the dozens of programs documented in Chapter 11 demonstrate that the alternative safety infrastructure can be built. Whether it gets built at the scale required to replace surveillance depends on whether the economic and political openings align.

The *bifurcated future* is the trajectory most consistent with the evidence in this book. Santa Cruz, Oakland, and Detroit have built real accountability structures. These wins are genuine. And simultaneously, the communities documented in Chapters 2, 6, and 7 — Detroit's Black neighborhoods, Minneapolis's East African communities, Hebron's Palestinian residents (see Chapter 12) — face infrastructure expanding and deepening regardless of what city councils and courts decide elsewhere. The geography of accountability follows the geography of political power. Winning in one city while losing in adjacent ones is not victory. It is the fracture line along which surveillance's contradictions are currently unresolved.

Shifting the Trajectory

The bifurcated future moves toward liberation when economic, technical, and political vulnerabilities align with organized community pressure at the same moment. Budget crises create openings — but only if communities have concrete alternatives ready when the surveillance contract comes up for renewal. Technical failures create political moments—but only if those failures are documented, publicized, and attached to identifiable decision-makers who can be held responsible. Legal victories create accountability surfaces—but only if communities have the organizational capacity to monitor compliance and enforce consequences.

The wildcard factors are the hardest to incorporate into any scenario: a major AI company's financial collapse; a catastrophic surveillance failure at a public event; climate-driven energy rationing that makes data center expansion politically impossible; a generation of voters who grew up under pervasive surveillance and decided, collectively, they did not want it. Angela Lipps was arrested at gunpoint in front of four children she was babysitting because a Clearview AI algorithm produced a match and no one in the chain of command thought to call her first (WDAY News, 2026). She is among more than a dozen Americans publicly documented as wrongfully arrested after facial recognition errors, with the ACLU tracking cases spanning 2019 through 2025 and noting that in every documented instance, police could have cleared the suspect before arrest through basic investigative steps (ACLU, 2025; Washington Post, 2025). The tenth case will come. Whether accumulation produces the political breaking point that ends systemic reliance on the technology—rather than one more settlement, followed by the same errors—depends on whether enough people are watching and refuse to stop watching.

Our Choice

On June 23, 2020, Justin Cummings stood before the Santa Cruz City Council and voted to ban the technology his city had helped create. He was direct about why. "We have technology that could target people of color in our community—it's technology we don't need. This is something that targets people like me" (KSBW, 2020). The vote was unanimous. A city that had incubated a predictive policing company decided it did not want what the company had built.

On April 9, 2026, Elon Musk's artificial intelligence company filed a 75-page complaint in federal court arguing that Colorado cannot require it to find out whether its AI system harms people differently based on race. The company asked the court to declare that accountability for algorithmic discrimination is unconstitutional. It argued that the national imperative of American AI dominance supersedes a state's authority to protect its residents from algorithmic harm. It declared the mechanics of its own system irrelevant to the question of whether that system can be regulated (X.AI LLC v. Weiser, 2026, Compl. paras. 13, 56).

These two moments—Cummings casting his vote and xAI filing its complaint—frame the actual choice this book has been building toward. Not a choice between technology and no technology. Not a choice between progress and reaction. The choice is between accountability and its absence. Between a surveillance infrastructure that communities can contest and one constitutionally insulated from contestation. Between systems that must answer for what they do to people and systems that have established, through contract law and trade secret protections and constitutional litigation, that what they do internally is no one else's business.

Robert Williams knows what the absence of accountability costs. His older daughter could not look at his picture for months after police arrested him in his driveway based on a facial recognition match they knew was unreliable (Williams,

2024). Porcha Woodruff, eight months pregnant, was arrested for a carjacking surveillance footage gave no reason to attribute to her (Michigan Public, 2024). Angela Lipps lost her home, her car, her dog, and five months of her life because a Clearview AI match functioned as institutional permission to stop investigating (Kolpack, 2026). Dnigma Howard was tackled down a staircase on her school's own security cameras and charged with felony aggravated battery for having a phone (Froelich, 2021). The data generated in each of these encounters persists. The surveillance systems that produced the encounters are still running.

What communities across this book demonstrate is that the choice is real. Santa Cruz chose. Oakland chose. The ACLU chose to litigate for three years on Robert Williams's behalf. Organizers chose to photograph cameras and map sensor locations and file public records requests and testify before city councils. Leaders of a Beautiful Struggle chose to sue Baltimore and won. These are not symbolic gestures. They are exercises of the same democratic authority that xAI is now asking federal courts to constitutionally constrain.

The book opened with Robert Williams standing in his driveway. The computer got it wrong. Then the system around that computer—the warrant, the arrest, the cell, the 30 hours—worked exactly as it was designed to work. That system is a choice. Not an inevitable product of technological development. Not a neutral outcome of data science. A choice made by specific companies, specific procurement officers, specific mayors, specific executives, and now specific lawyers filing constitutional claims in federal courts. Built by people. Which means something else is also possible.

The communities in this book are building it. The question is whether the legal architecture being constructed to stop them will succeed—and whether enough people understand what is being decided, in courtrooms and city councils and procurement offices, while the cameras keep running and the algorithms keep scoring and the surveillance infrastructure keeps expanding into every space where people have not yet organized to push it back.

The lens is watching. The question is whether we are watching back.

Chapter 14: The System Cannot Be Repaired—It Must Be Rebuilt

When I began researching this book, I approached it as a scholar documenting how technology reproduces inequality. That is where the argument starts—in Chapter 1, with Robert Williams arrested in his driveway, the system working exactly as designed. By Chapter 11, the argument had shifted. The question was no longer how to make the system more fair. The question was whether it should exist at all.

The pushback has been consistent: that I am centering race divisively, dismissing legitimate safety concerns, or proposing abolition that will not work. Why not regulate surveillance better, control its harms, or build in oversight?

This chapter is my answer. Not a defense—a clarification.

The current surveillance system cannot be reformed into something just. It was built to do something specific. That something is racial control. Better regulations do not change the purpose. They change the paperwork.

What Regulation Leaves Untouched

The managed surveillance state—the future in which CCOPS ordinances govern acquisition, transparency requirements illuminate operations, and city councils vote annually on surveillance spending—is more just than the current system. It is not just enough.

Detroit police agreed in June 2024 to require independent verification of facial recognition matches before arrest. That is regulation. It is also real. The settlement was one of the strongest in the country. And yet Detroit police still use facial recognition. The technology remains. The infrastructure persists. What changed is the process.

Process matters. A community that can contest surveillance acquisition, demand accountability when the system fails, and vote to defund failed systems has power it did not have before. That is worth defending.

But regulation leaves the fundamental architecture intact. The same databases that produce wrongful arrests continue. The same algorithms that misidentify Black faces at disproportionate rates continue. The same neighborhoods continue being monitored more densely. The system that produced Angela Lipps's arrest, Porcha Woodruff's seven-month detention while pregnant, and the surveillance of schoolchildren in predominantly Black districts continues—now with a city council oversight meeting attached.

San Francisco banned facial recognition in 2019. In 2024, SFPD violated the ban at least twelve times by asking neighboring agencies to run searches on its behalf. The ban created accountability surfaces. It did not prevent the behavior. When accountability failed, San Francisco voters passed Proposition E in March 2024, expanding police surveillance authority despite documented violations of the previous year's regulations. One city's regulation became political terrain. The infrastructure adapted. Police obtained what they wanted through a different mechanism.

Regulation is insufficient because it operates within the assumption that surveillance of this kind is legitimate. It is not.

What the Current System Actually Is

The current surveillance infrastructure is not a neutral tool applied unfairly. It is a system designed, built, and deployed to concentrate monitoring and control on specific communities. That design predates facial recognition by centuries.

Chapter 1 traces that genealogy: slave patrols, Jim Crow biometrics, War on Drugs militarization. Each era claimed technological advancement toward objectivity. Each deployed that technology first and most thoroughly against Black communities. The pattern is not accidental. It is the system working as designed.

The architects of that design include police departments that purchased facial recognition despite documented failure rates 10–100 times higher for Black faces; prosecutors who acted on algorithmic matches without verification; city governments that installed cameras in redlined neighborhoods while wealthier areas remained unwatched; federal agencies that built vendor lock-in so deep that replacing systems would cost $100 million; and technology companies that trained algorithms on datasets overwhelmingly composed of lighter-skinned faces and marketed them as neutral.

These are choices—not inevitable technical developments. Regulating the system does not change what it is. It adds process to a fundamentally unjust architecture.

Why Reconstruction, Not Just Removal

If the problem were simply that surveillance exists, the answer would be simpler. But the problem is that surveillance as currently deployed *replaces* the public investment communities need to be safe.

Chicago spent $50 million on ShotSpotter by September 2024. Detroit spent $7 million on ShotSpotter expansion in 2022. DHS distributed $615 million in Urban Area Security Initiative funding in 2022, much of it used for surveillance technology.

What if that money had gone to mental health crisis response, housing, youth employment, or community health centers?

The answer is documented.

Denver's STAR program operates for $7.2 million annually and responds to 7,000 calls—46% of eligible calls citywide. The average cost per response is $151. When those same calls go through traditional police processing, the cost reaches $646 per person. STAR reduces low-level crime by 34% without increasing serious crime.

CAHOOTS operated for 35 years in Eugene, Oregon. A medic and a crisis worker responded to welfare checks, conflicts, and crises—handling between 5% and 23% of emergency calls depending on methodology. The program cost $820,000 under its last contract. The Eugene Police Department costs $68 million.

These are not hypothetical alternatives. They are operating programs. Reconstruction means scaling them. It means redirecting surveillance budgets toward community safety infrastructure that actually works.

Dismantling the current system is necessary not because removal is the goal, but because the money tied up in surveillance is the same money needed to build something better.

What Dismantling Actually Means

This is concrete, not theoretical.

Facial recognition systems would be decommissioned. Databases deleted. Contracts terminated. NIST testing ended. Police would not have access to these tools because the tools would not exist.

Predictive policing algorithms—PredPol, HunchLab, COMPAS—would be dismantled. Proprietary algorithms destroyed. Contracts terminated. Federal funding stopped.

ShotSpotter and similar acoustic surveillance networks would be decommissioned. Microphones removed. Data deleted. Contracts ended. Cities have already done this—Houston, Baton Rouge, Atlanta, Chicago—proving it is possible.

License plate readers, gang databases, school resource officers with facial recognition access, and integration platforms that bundle these streams would be dismantled.

The decision to monitor people would be made differently—not through automated systems operating at scale and speed, but through human judgment, community accountability, and legal processes that allow contestation.

This requires sustained organizing, legal victories, and political will. But it is not speculative. Cities are already doing parts of it.

What We Build Instead

The current system exists not because we need surveillance, but because we defunded everything else.

In neighborhoods where ShotSpotter operates—Chicago's South and West Sides, East Baltimore, South Los Angeles—mental health services are scarce, schools are underfunded, and economic investment has been withheld for decades. Surveillance fills the space created by disinvestment. It does not solve it. It manages the people produced by it.

Reconstruction means building different infrastructure: community-based crisis response; mental health services accessible without police involvement; housing investment tied to safety; economic opportunity; and restorative justice instead of criminal justice.

These are not soft alternatives. They are systems with evidence bases, operating budgets, and professional workforces. Reconstruction means scaling them.

The Timeline Is Real

This cannot happen overnight. San Francisco passed a facial recognition ban in 2019. In 2024, it was nearly overturned. Bans can be reversed. Surveillance spending can increase. The political battle is continuous.

But the surveillance system has structural vulnerabilities: economic fragility, technical brittleness, and energy demands that will eventually compete with basic public services. These vulnerabilities create openings—budget crises, technical failures, political moments. Each has produced surveillance contract cancellations.

When those moments arrive, the question is whether communities have alternatives ready.

What I Believe

I believe the current surveillance infrastructure should be dismantled—not reformed, not regulated better, but dismantled, decommissioned, and deleted.

I believe it is possible. Communities are already doing parts of it.

I believe it is necessary. The evidence shows the system concentrates harm on Black communities, produces measurable psychological injury, and cannot be reformed into justice because that is not its purpose.

I do not believe it is immediately achievable everywhere. But I believe the choice is real. Communities are making it. Santa Cruz chose to dismantle what it had built. Oakland chose to reject predictive policing. Detroit residents chose to sue.

When moments of choice arrive—budget crises, court rulings, shifts in political possibility—different decisions become available.

The lens is watching. The question is not whether we can reform it to be kinder. The question is whether we can dismantle it and build something else.

That work is already happening. The communities in this book show what is possible.

References

Adamson, C. R. (1983). Punishment after slavery: Southern state penal systems, 1865–1890. Social Problems, 30(5), 555–569.

Action Center on Race and the Economy. (2022, December). DHS funds surveillance technology in US cities [Report]. https://acrecampaigns.org/press-release/dhs-funds-surveillance-technology-in-us-cities-report-says/

Al Jazeera. (2022, January 20). Facial recognition taken to court in India's surveillance hotspot. Al Jazeera. https://www.aljazeera.com/news/2022/1/20/india-surveillance-hotspot-telangana-facial-recognition-court-lawsuit-privacy

Alang, S., VanHook, C., Judson, J., Ikiroma, A., & Adkins-Jackson, P. B. (2022). Police brutality, heightened vigilance, and the mental health of Black adults. Psychology of Violence, 12(4), 211–220. https://doi.org/10.1037/vio0000418

Alexander, C. (2026, March 25). Lipps' attorneys respond to Fargo PD press conference. Valley News Live. https://www.valleynewslive.com/2026/03/24/lipps-attorneys-respond-fargo-pd-press-conference-basic-investigative-efforts-were-never-made/

Allen, A. L. (2022). Dismantling the "Black Opticon:" Privacy, race equity, and online data-protection reform. Yale Law Journal Forum. https://doi.org/10.2139/ssrn.4027234

Almeida, D., Shmarko, K., & Lomas, E. (2022). The ethics of facial recognition technologies, surveillance, and accountability in an age of artificial intelligence: A comparative analysis of US, EU, and UK regulatory frameworks. AI and Ethics, 2, 377–387. https://doi.org/10.1007/s43681-021-00077-w

American Civil Liberties Union. (2018, July 26). Amazon's face recognition falsely matched 28 members of Congress with mugshots [Press release]. https://www.aclu.org/news/privacy-technology/amazons-face-recognition-falsely-matched-28

American Civil Liberties Union. (2020). How is face recognition surveillance technology racist? https://www.aclu.org/news/privacy-technology/how-is-face-recognition-surveillance-technology-racist

American Civil Liberties Union. (2024a, June 28). Civil rights advocates achieve the nation's strongest police department policy on facial recognition technology [Press release]. https://www.aclu.org/press-releases/civil-rights-advocates-achieve-the-nations-strongest-police-department-policy-on-facial-recognition-technology

American Civil Liberties Union. (2024b). Williams v. City of Detroit: Face recognition false arrest. https://www.aclu.org/cases/williams-v-city-of-detroit-face-recognition-false-arrest

American Civil Liberties Union. (2026). More than a dozen wrongful arrests due to police reliance on facial recognition technology. https://www.aclu.org/news/privacy-technology/more-than-a-dozen-wrongful-arrests-due-to-police-reliance-on-facial-recognition-technology

American Civil Liberties Union of New Jersey. (2024, January 26). ACLU-NJ and ACLU national file amicus in a challenge to a wrongful arrest due to face recognition [Press release]. https://www.aclu-nj.org/press-releases/aclu-nj-and-aclu-national-file-amicus-challenge-wrongful-arrest-due-face-recognition/

Amnesty International. (2023, May 2). Automated apartheid: How facial recognition fragments, segregates and controls Palestinians in the OPT. https://www.amnesty.org/en/latest/news/2023/05/israel-opt-israeli-authorities-are-using-facial-recognition-technology-to-entrench-apartheid/

Amnesty International India. (n.d.). Ban the Scan Hyderabad. https://banthescan.amnesty.org/hyderabad/index.html

Angwin, J., Larson, J., Mattu, S., & Kirchner, L. (2016, May 23). Machine bias. ProPublica. https://www.propublica.org/article/machine-bias-risk-assessments-in-criminal-sentencing

Balko, R. (2013). Rise of the warrior cop: The militarization of America's police forces. PublicAffairs.

Ballotpedia. (2024). San Francisco, California, Proposition E, Limit Police Department Administrative Task Time and Increase Use of Camera and Drone Technology Initiative (March 2024). https://ballotpedia.org/San_Francisco,_California,_Proposition_E,_Limit_Police_Department_Administrative_Task_Time_and_Increase_Use_of_Camera_and_Drone_Technology_Initiative_(March_2024)

Balto, S. (2019). Occupied territory: Policing Black Chicago from Red Summer to Black Power. University of North Carolina Press.

Ban the Scan. (n.d.). Ban the Scan campaign. https://www.banthescan.org/

Bass, S. (2001). Policing space, policing race: Social control imperatives and police discretionary decisions. Social Justice, 28(1), 156–176.

Benjamin, R. (2019). Race after technology: Abolitionist tools for the new Jim code. Polity Press.

Benson, B. L., Rasmussen, D. W., & Sollars, D. L. (1995). Police bureaucracies, their incentives, and the war on drugs. Public Choice, 83(1–2), 21–45.

Booker, M. D. (2022). Empowering, engaging, and equipping technology through acceptance: A quantitative study utilizing the Modified Technology Acceptance Model (mTAM) to explore facial-recognition software acceptance (FRSA) in the Smart Policing Initiative (SPI) [Doctoral dissertation]. https://www.researchgate.net/publication/393441006

Bor, J., Venkataramani, A. S., Williams, D. R., & Tsai, A. C. (2018). Police killings and their spillover effects on the mental health of Black Americans: A population-based, quasi-experimental study. Lancet, 392(10144), 302–310. https://doi.org/10.1016/S0140-6736(18)31130-9

Brown, L. K. (2005). Officer or overseer? Why police desegregation fails as an adequate solution to racist, oppressive, and violent policing in Black communities. NYU Review of Law and Social Change, 29, 757–795.

Brown, T. B., Mané, D., Roy, A., Abadi, M., & Gilmer, J. (2017). Adversarial patch. arXiv preprint arXiv:1712.09665. https://doi.org/10.48550/arXiv.1712.09665

Buolamwini, J., & Gebru, T. (2018). Gender shades: Intersectional accuracy disparities in commercial gender classification. Proceedings of Machine Learning Research, 81, 77–91.

Bulletin of the Atomic Scientists. (2022, October 27). China's high-tech surveillance drives oppression of Uyghurs. Bulletin of the Atomic Scientists. https://thebulletin.org/2022/10/chinas-high-tech-surveillance-drives-oppression-of-uyghurs/

Bureau of Justice Statistics. (2008). Drugs and crime facts. U.S. Department of Justice.

Bureau of Justice Statistics. (2011). Census of state and local law enforcement agencies, 2008. U.S. Department of Justice.

Cameron, D. (2025). Here's the company that sold DHS ICE's notorious face recognition app. WIRED. https://www.wired.com/story/mobile-fortify-ice-dhs-nec-face-recognition/

Cameron, D., & Varner, M. (2026, February 5). ICE and CBP's face-recognition app can't actually verify who people are. WIRED. https://www.wired.com/story/ice-cbp-facial-recognition-mobile-fortify/

Caramanica, J., Morris, W., & Wortham, J. (2016, February 6). Beyoncé in 'Formation': Entertainer, activist, both? The New York Times. https://www.nytimes.com/2016/02/07/arts/music/beyonce-formation-super-bowl-black-lives-matter.html

CBC News. (2019, August 11). Hong Kong protesters use laser pointers to deter police, scramble facial recognition. CBC News. https://www.cbc.ca/news/world/hong-kong-protest-lasers-facial-recognition-technology-1.5240651

Chicago Office of Inspector General. (2021, August 24). The Chicago Police Department's use of ShotSpotter technology. https://igchicago.org/publications/cpd-use-of-shotspotter-technology/

City of Atlanta, Department of Atlanta Information Management. (2020). Notification of decision to discontinue ShotSpotter application [Internal memorandum].

Collier, K., Perlo, J., Lee, J., & Lavietes, M. (2026, February 6). How ICE agents are using facial recognition technology to bring surveillance to the streets. NBC News. https://www.nbcnews.com/tech/security/ice-agent-facial-recognition-video-protest-movile-fortify-photo-rcna257331

Columbia Law Review. (2025). Police technology experiments. Columbia Law Review, 125(1). https://columbialawreview.org/content/police-technology-experiments/

Cook, D. (2017). Policing the black underclass: The untold story of broken windows policing and the order maintenance police reform movement, 1967–1994 [Honors thesis, University of Virginia].

Cooper, H. L. F., Moore, L., Gruskin, S., & Krieger, N. (2015). The impact of a police drug crackdown on drug injectors' ability to practice harm reduction: A qualitative study. Social Science & Medicine, 61(3), 673–684.

Cornell Chronicle. (2025, November). Roadmap shows the environmental impact of AI data center boom. Cornell University. https://news.cornell.edu/stories/2025/11/roadmap-shows-environmental-impact-ai-data-center-boom

Cowper, T. J. (2004). Improving the view of the world: Law enforcement and augmented reality technology. FBI Law Enforcement Bulletin, 73(1), 12–18.

Cronin, T. W., Cronin, T. Z., & Millakovich, M. E. (1981). U.S. v. crime in the streets. Indiana University Press.

Dee, T. S., & Pyne, J. (2022). A community response approach to mental health and substance abuse crises reduced crime. Science Advances, 8(23), eabm2106. https://doi.org/10.1126/sciadv.abm2106

Democracy Now! (2021, March 5). "We Do This 'Til We Free Us": Mariame Kaba on abolishing police, prisons & moving toward justice [Video interview]. https://www.democracynow.org/2021/3/5/george_floyd_policing_act_mariame_kaba

Detroit News. (2022, October 11). Detroit City Council OKs $7 million ShotSpotter expansion. The Detroit News. https://eu.detroitnews.com/story/news/local/detroit-city/2022/10/11/detroit-city-council-oks-shotspotter-7-million-expansion/69552267007/

Driver, A., & Jean-Baptiste, R. (2024, September 20). Notice of intent to sole source NEC Advanced Identity Manager. SAM.gov. https://sam.gov/opp/b016354c5bd045fa92e4886878747dc8/view

Electronic Frontier Foundation. (2021, January 14). Oakland's progressive fight to protect residents from government surveillance. EFF Deeplinks. https://www.eff.org/deeplinks/2021/01/oaklands-progressive-fight-protect-residents-government-surveillance

Electronic Frontier Foundation. (2024, December 26). Police surveillance in San Francisco: 2024 in review. EFF Deeplinks. https://www.eff.org/deeplinks/2024/12/police-surveillance-san-francisco-2024-year-review

Electronic Frontier Foundation. (2025, December 28). Procurement power — when cities realized they can just say no: 2025 in review. EFF Deeplinks. https://www.eff.org/deeplinks/2025/12/procurement-power-when-cities-realized-they-can-just-say-no-2025-review

Eugene Police Department Crime Analysis Unit. (2019). CAHOOTS program analysis. City of Eugene. https://www.eugene-or.gov/DocumentCenter/View/56717/CAHOOTS-Program-Analysis

Feeley, M., & Sarat, A. (1980). The policy dilemma: Federal crime policy and the Law Enforcement Assistance Administration, 1968–1978. University of Minnesota Press.

Felker-Kantor, M. (2018). Policing Los Angeles: Race, resistance, and the rise of the LAPD. University of North Carolina Press.

Ferguson, A. G. (2025). Video analytics and Fourth Amendment vision. Texas Law Review. https://digitalcommons.wcl.american.edu/facsch_lawrev/2256

Froelich, A. (2021). The increasingly covert and invasive surveillance of students and its visible role in the student-to-prison pipeline. Children's Legal Rights Journal, 40(2), 118–142. https://lawecommons.luc.edu/clrj/vol40/iss2/4

Garvie, C., Bedoya, A. M., & Frankle, J. (2016, October 18). The perpetual line-up: Unregulated police face recognition in America. Georgetown Law Center on Privacy & Technology. https://www.perpetuallineup.org

Gonzalez, S. (2021). Slave patrols and the history of policing in America. Insights, Spring 2021, 1–4.

Governing. (2021, April 21). Santa Cruz, Calif., becomes first to ban predictive policing. https://www.governing.com/security/santa-cruz-calif-becomes-first-to-ban-predictive-policing.html

Grother, P. (2022). Face recognition vendor test (FRVT) Part 8: Summarizing demographic differentials (NIST Interagency Report 8429). National Institute of Standards and Technology. https://doi.org/10.6028/NIST.IR.8429

Grother, P., Ngan, M., & Hanaoka, K. (2019). Face recognition vendor test (FRVT) Part 3: Demographic effects (NIST Interagency Report 8280). National Institute of Standards and Technology. https://doi.org/10.6028/NIST.IR.8280

Hadden, S. E. (2003). Slave patrols: Law and violence in Virginia and the Carolinas. Harvard University Press.

Hart, R. (2024, September 3). Clearview AI—Controversial facial recognition firm—fined $33 million for "illegal database." Forbes. https://www.forbes.com/sites/roberthart/2024/09/03/clearview-ai-controversial-facial-recognition-firm-fined-33-million-for-illegal-database/

Harvey, A. (2010). CV Dazzle: Camouflage from face detection. https://cvdazzle.com/

Hendrix, J. A., Taniguchi, T. A., Strom, K. J., Barrick, K., & Johnson, N. J. (2018). The eyes of law enforcement in the new panopticon: Police-community racial asymmetry and the use of surveillance technology. Surveillance & Society, 16(1), 53–68. https://ojs.library.queensu.ca/index.php/surveillance-and-society/index

Herskind, M. (2021, March 1). Some lessons from Mariame Kaba's "We Do This 'Til We Free Us" [Blog post]. Medium. https://micahherskind.medium.com/some-lessons-from-mariame-kabas-we-do-this-til-we-free-us-4ae4fc9986bb

Hinton, E., & Cook, D. (2021). The mass criminalization of Black Americans: A historical overview. Annual Review of Criminology, 4, 261–286. https://doi.org/10.1146/annurev-criminol-060520-033306

Hubbard, S. (2026, February 6). Mission creep: AI surveillance at DHS crosses dangerous line into tracking Americans. American Immigration Council. https://www.americanimmigrationcouncil.org/research/ai-surveillance-dhs-mission-creep

Ingram, G. (2021). The applicability of understanding hypervigilance in law enforcement [Senior honors thesis, Liberty University].

Institute for Middle East Understanding. (2021). Fact sheet: Israeli surveillance & restrictions on Palestinian movement.

https://imeu.org/resources/resources/fact-sheet-israeli-surveillance-restrictions-on-palestinian-movement/128

The Intercept. (2022, September 17). Detroit may put $7 million in Covid aid to police surveillance. The Intercept. https://theintercept.com/2022/09/17/police-surveillance-shotspotter-detroit/

Kaba, M. (2021). We do this 'til we free us: Abolitionist organizing and transforming justice. Haymarket Books.

Khoo, C. (2022). Facial recognition technology and the growing power of artificial intelligence [Testimony]. Standing Committee on Access to Information, Privacy and Ethics (ETHI), House of Commons, Canada.

Kolpack, D. (2026, March 14). AI error jails innocent grandmother for months in Fargo case. InForum. https://www.inforum.com/news/fargo/ai-error-jails-innocent-grandmother-for-months-in-fargo-case

KSBW. (2020, December 11). Santa Cruz to become first city to ban predictive policing. https://www.ksbw.com/article/santa-cruz-to-become-first-city-to-ban-predictive-policing/32976124

KVAL. (2025, December). When reform backfires: How the CAHOOTS Act helped dismantle Eugene's mobile crisis program. KVAL News. https://kval.com/news/local/when-reform-backfires-how-the-cahoots-act-helped-dismantle-eugenes-mobile-crisis-program

Lateral Reviews. (2021). Review of We Do This 'Til We Free Us by Mariame Kaba. Lateral: Journal of the Cultural Studies Association. https://csalateral.org/reviews/we-do-this-til-we-free-us-kaba-kelly/

Lau, T. (2020). Predictive policing explained. Brennan Center for Justice. https://www.brennancenter.org/our-work/research-reports/predictive-policing-explained

Law Quadrangle. (2024, Winter). Flawed facial recognition technology leads to wrongful arrest and historic settlement.

https://quadrangle.michigan.law.umich.edu/issues/winter-2024-2025/flawed-facial-recognition-technology-leads-wrongful-arrest-and-historic

Leaders of a Beautiful Struggle v. Baltimore Police Department, 2 F.4th 330 (4th Cir. 2021).

Liberty. (2020, August 11). Legal challenge: Ed Bridges v South Wales Police. https://www.libertyhumanrights.org.uk/issue/legal-challenge-ed-bridges-v-south-wales-police/

Los Angeles Police Commission. (2019). Review of selected Los Angeles Police Department data-driven policing strategies. Office of the Inspector General.

Lum, K., & Isaac, W. (2016). To predict and serve? Significance, 13(5), 14–19. https://doi.org/10.1111/j.1740-9713.2016.00960.x

Lynch, M. (2012). Theorizing the role of the "war on drugs" in US punishment. Theoretical Criminology, 16(2), 175–199.

Marshall Project. (2024, July 25). Sending unarmed responders instead of police: What we've learned. The Marshall Project. https://www.themarshallproject.org/2024/07/25/police-mental-health-alternative-911

Martín Ríos, P. (2024). Predictive algorithms and criminal justice: Expectations, challenges and a particular view of the Spanish VioGén system. Rivista Italiana di Informatica e Diritto, 2, 547–565.

MediaJustice, Mijente, & Immigrant Defense Project. (2020). Unmasked: Impacts of pandemic policing. https://mediajustice.org/wp-content/uploads/2020/10/Unmasked.pdf

Mehrotra, D., Mattu, S., Gilbertson, A., & Sankin, A. (2021, December 2). How we determined predictive policing software disproportionately targeted low-income, Black, and Latino neighborhoods. The Markup. https://themarkup.org/prediction-bias/2021/12/02/crime-prediction-software-promised-to-be-free-of-biases-new-data-shows-it-perpetuates-them

Metro Times. (2024, June 28). Detroit police to overhaul facial recognition use after groundbreaking settlement. https://www.metrotimes.com/news/detroit-police-to-overhaul-facial-recognition-use-after-groundbreaking-settlement-in-false-arrest-suit-36657192

Michigan Public. (2024, June 28). "It didn't make sense at all": Wrongful facial recognition arrest in Detroit leads to landmark settlement. Michigan Public. https://www.michiganpublic.org/criminal-justice-legal-system/2024-06-28/it-didnt-make-sense-at-all-wrongful-facial-recognition-arrest-leads-to-landmark-settlement

Michigan Public Radio. (2022, October 11). Detroit City Council approves $7 million ShotSpotter expansion. Michigan Public Radio. https://www.michiganpublic.org/public-safety/2022-10-11/detroit-city-council-approves-7-million-shotspotter-expansion

MIT Technology Review. (2021, April 14). The new lawsuit that shows facial recognition is officially a civil rights issue. https://www.technologyreview.com/2021/04/14/1022676/robert-williams-facial-recognition-lawsuit-aclu-detroit-police/

MIT Technology Review. (2022, December 7). US police use counterterrorism money to buy spy tech. MIT Technology Review. https://www.technologyreview.com/2022/12/07/1064354/police-counterterrorism-money-buy-spy-tech-surveillance-fema/

Mohler, G. O., Short, M. B., Brantingham, P. J., Schoenberg, F. P., & Tita, G. E. (2011). Self-exciting point process modeling of crime. Journal of the American Statistical Association, 106(493), 100–108.

Murakawa, N. (2014). The first civil right: How liberals built prison America. Oxford University Press.

National Association for the Advancement of Colored People & Mississippi State Conference of the NAACP v. X.AI Corp. & MZX Tech LLC, Civil Case No. 3:26-cv-00074-MPM-JMV (U.S.

District Court for the Northern District of Mississippi, filed April 14, 2026).

National Institute of Standards and Technology. (2026, January 23). Face recognition technology evaluation (FRTE) 1:1 verification [Data tables and summary]. https://pages.nist.gov/frvt/html/frvt11.html

National League of Cities. (2024a). Denver, CO: Community response model. https://www.nlc.org/resource/reimagining-public-safety-impact-updates/denver-co-community-response-model/

National League of Cities. (2024b). Eugene, OR: Community response model. https://www.nlc.org/resource/reimagining-public-safety-impact-updates/eugene-or-community-response-model/

NBC News. (2020, December 29). Black man in New Jersey misidentified by facial recognition tech and falsely jailed, lawsuit claims. https://www.nbcnews.com/news/us-news/black-man-new-jersey-misidentified-facial-recognition-tech-falsely-jailed-n1252489

NBC News. (2021, February 23). Black History Month: Mariame Kaba wants us to imagine a future without prisons. NBC News. https://www.nbcnews.com/news/nbcblk/black-history-month-mariame-kaba-wants-us-imagine-future-without-prisons-n1258586

Nessa, M. (2026, March 14). Michael Nessa for Angela Robin Lipps [Fundraiser]. GoFundMe. https://www.gofundme.com/f/michael-nessa-for-angela-robin-lipps

NPR. (2020, June 24). 'The computer got it wrong': How facial recognition led to false arrest of Black man. https://www.npr.org/2020/06/24/882683463/the-computer-got-it-wrong-how-facial-recognition-led-to-a-false-arrest-in-michig

Nunn, K. B. (2002). Race, crime and the pool of surplus criminality: Or why the war on drugs was a war on Blacks. Journal of Gender, Race and Justice, 6, 381–445.

Office of the United Nations High Commissioner for Human Rights. (2022, August 31). OHCHR assessment of human rights concerns in the Xinjiang Uyghur Autonomous Region, People's Republic of China. https://www.ohchr.org/en/documents/country-reports/ohchr-assessment-human-rights-concerns-xinjiang-uyghur-autonomous-region

Pagan, N., Baumann, J., Elokda, E., De Pasquale, G., Bolognani, S., & Hannák, A. (2023). A classification of feedback loops and their relation to biases in automated decision-making systems. In Proceedings of the 3rd ACM Conference on Equity and Access in Algorithms, Mechanisms, and Optimization (Article 7, pp. 1–14). Association for Computing Machinery. https://doi.org/10.1145/3617694.3623227

Participedia. (2019). Hong Kong protesters implement methods to avoid facial recognition technology and government tracking. https://participedia.net/case/hong-kong-protestors-implement-methods-to-avoid-facial-recognition-technology-and-government-tracking

Pew Research Center. (2025, October 24). What we know about energy use at U.S. data centers amid the AI boom. Pew Research Center. https://www.pewresearch.org/short-reads/2025/10/24/what-we-know-about-energy-use-at-us-data-centers-amid-the-ai-boom/

R (Bridges) v. Chief Constable of South Wales Police [2020] EWCA Civ 1058. Court of Appeal of England and Wales.

Radio Free Asia. (2021, August 5). US tech products enable Chinese surveillance in Xinjiang, researchers find. Radio Free Asia. https://www.rfa.org/english/news/uyghur/us-tech-products-08052021185345.html

Richardson, R., Schultz, J., & Crawford, K. (2019). Dirty data, bad predictions: How civil rights violations impact police data, predictive policing systems, and justice. New York University Law Review Online, 94, 192–233.

San Francisco Board of Supervisors. (2019). Administrative code: Acquisition of surveillance technology (Ordinance No. 190110). https://www.sanfranciscopolice.org/your-sfpd/policies/19b-surveillance-technology-policies

San Francisco Department of Elections. (2024, March 22). March 5, 2024 final election results: Summary report [Certified]. https://www.sfelections.org/results/20240305/data/20240322/summary.pdf

Sankin, A., & Mattu, S. (2023, October 2). Predictive policing software terrible at predicting crimes. The Markup. https://themarkup.org/prediction-bias/2023/10/02/predictive-policing-software-terrible-at-predicting-crimes

Santa Cruz Works. (2020, June 25). Santa Cruz is the first city to ban predictive policing. https://www.santacruzworks.org/news/santa-cruz-is-the-first-city-to-city-ban-predictive-policing

Santoshini, S. (2022, October 11). Indian police use facial recognition to persecute Muslims and other marginalized communities. Coda Story. https://www.codastory.com/authoritarian-tech/india-police-facial-recognition/

Saunders, J., Hunt, P., & Hollywood, J. S. (2016). Predictions put into practice: A quasi-experimental evaluation of Chicago's predictive policing pilot. Journal of Experimental Criminology, 12, 347–371. https://doi.org/10.1007/s11292-016-9272-0

Saunders, S. M. (2020). Surveilling modern surveillance: An examination of the impacts of surveillance on marginalized identities and police behaviour [Master's thesis, York University].

Secure Justice. (2021, January 15). First-in-the-nation surveillance tech bans. https://secure-justice.org/blog/first-in-the-nation-bans-predictive-analytics-biometric-surveillance-technologies

SF Standard. (2024, July 18). SFPD skirted facial-recognition ban, lawsuit says. https://sfstandard.com/2024/07/18/san-francisco-police-facial-recognition-violations/

Sharma, M. K. (n.d.). To be, or not to be? Regulating impossible AI in the United States [Unpublished manuscript].

Smith, N. A., Voisin, D. R., Yang, J. P., & Tung, E. L. (2019). Keeping your guard up: Hypervigilance among urban residents affected by community and police violence. Health Affairs, 38(10), 1662–1669. https://doi.org/10.1377/hlthaff.2019.00560

Somerville, H. (2026, February 3). Palantir achieves another revenue record with $1.41 billion quarter. The Wall Street Journal.

Sottile, Z. (2026, March 29). Police used AI facial recognition to arrest a Tennessee woman for crimes committed in a state she says she's never visited. CNN. https://www.cnn.com/2026/03/29/us/angela-lipps-ai-facial-recognition

Stateline. (2025, February 4). Facial recognition in policing is getting state-by-state guardrails. https://stateline.org/2025/02/04/facial-recognition-in-policing-is-getting-state-by-state-guardrails/

StateScoop. (2024, May 31). Houston mayor calls gunshot detection tech a 'gimmick,' plans to end ShotSpotter contract. StateScoop. https://statescoop.com/houston-shotspotter-gimmick/

StateScoop. (2025, February 12). Baton Rouge police department cancels ShotSpotter contract. StateScoop. https://statescoop.com/baton-rouge-police-department-cancels-contract-shotspotter-2025/

Stewart, M. (2025). Beyond explanation: Evidentiary rights for algorithmic accountability. arXiv preprint arXiv:2603.22716. https://doi.org/10.48550/arXiv.2603.22716

TechCrunch. (2019, May 14). San Francisco passes city government ban on facial recognition tech. https://techcrunch.com/2019/05/14/san-francisco-facial-recognition-ban/

The Appeal. (2024, October). Cities might ditch ShotSpotter, but not police technology. The Appeal. https://theappeal.org/shotspotter-police-technology/

The Nation. (2025, June 24). This organization proved there was an alternative to the police. Now it's being defunded. The Nation. https://www.thenation.com/article/society/cahoots-crisis-response-police-alternatives/

Thompson, H. A. (2004). Whose Detroit? Politics, labor, and race in a modern American city. Cornell University Press.

Tipton, J. L., Sirotin, Y. B., Howard, J. J., & Vemury, A. R. (2021). Quantifying the extent to which race and gender features determine identity in commercial face recognition algorithms (DHS Science and Technology Directorate Technical Paper Series). Department of Homeland Security. https://www.dhs.gov/sites/default/files/publications/21_0922_st_quantifying-commercial-face-recognition-gender-and-race_updated.pdf

Tonry, M. (1994). Malign neglect: Race, crime, and punishment in America. Oxford University Press.

UAB Institute for Human Rights. (2025, February 13). The abuse of facial recognition technology in the Hong Kong protests. UAB Institute for Human Rights Blog. https://sites.uab.edu/humanrights/2025/02/13/the-abuse-of-facial-recognition-technology-in-the-hong-kong-protests/

Ulbricht, L., & Egbert, S. (2024). In Palantir we trust? Regulation of data analysis platforms in public security. Big Data & Society, 11(3), 1–15. https://doi.org/10.1177/20539517241255108

University of Chicago Crime Lab. (2024). Strategic Decision Support Centers (SDSCs). https://crimelab.uchicago.edu/projects/strategic-decision-support-centers-sdscs/

University of Oregon Social Sciences. (2025, April 15). Researchers present findings in wake of CAHOOTS losing funding. University of Oregon. https://socialsciences.uoregon.edu/researchers-present-findings-wake-cahoots-losing-funding

Wani, M. N. (n.d.). Predictive policing and the ghost in the machine: AI, crime, and the collapse of presumption [Unpublished manuscript].

Washington Post. (2021, November 8). Israel escalates surveillance of Palestinians with facial recognition program in West Bank. The Washington Post. https://www.washingtonpost.com/world/middle_east/israel-palestinians-surveillance-facial-recognition/2021/11/05/3787bf42-26b2-11ec-8739-5cb6aba30a30_story.html

Washington Post. (2025, January 13). Arrested by AI: Police ignore standards after facial recognition matches. The Washington Post. https://www.washingtonpost.com/business/interactive/2025/police-artificial-intelligence-facial-recognition/

WDAY News. (2026, March 11). AI error jails innocent grandmother for months in Fargo fraud case [Video report]. WDAY News.

Weaver, V. M. (2007). Frontlash: Race and the development of punitive crime policy. Studies in American Political Development, 21(2), 230–265.

Webb, E. (2025, February 21). CEO of Clearview AI, the startup that scraped billions of online face images, resigns. Business Insider. https://www.businessinsider.com/clearview-ai-ceo-resigns-2025

White Bird Clinic. (n.d.). What is CAHOOTS? https://whitebirdclinic.org/what-is-cahoots/

Williams, R. (2021, July 13). Michigan man wrongfully accused with facial recognition urges Congress to act. Detroit News. https://www.detroitnews.com/story/news/politics/2021/07/13/house-panel-hear-michigan-man-wrongfully-accused-facial-recognition/7948908002/

Williams, R. (2023, February 24). I did nothing wrong. I was arrested anyway. American Civil Liberties Union. https://www.aclu.org/news/privacy-technology/i-did-nothing-wrong-i-was-arrested-anyway

The Wire. (2025, July 2). As AI took over policing in Delhi, who bore the brunt? The Wire. https://m.thewire.in/article/government/delhi-police-ai-facial-recognition

World Economic Forum. (2019, May 14). San Francisco votes to ban city use of facial recognition technology. https://www.weforum.org/stories/2019/05/san-francisco-votes-to-ban-city-use-of-facial-recognition-technology/

X.AI LLC v. Weiser, No. 1:26-cv-01515 (D. Colo. filed Apr. 9, 2026) (Complaint for Declaratory and Injunctive Relief).

Yen, E. (2021). Predictive policing and its ramifications: The use of AI in the justice system and the future of law enforcement [Student paper]. Binghamton University.

Young-Drake, R., Aladetimi, T., Chambers, A., & Radford, B. (n.d.). A growing dilemma: How police brutality affects mental health in Black communities [Unpublished manuscript]. Howard University School of Law, Movement Lawyering Clinic.

www.ingramcontent.com/pod-product-compliance
Lightning Source LLC
LaVergne TN
LVHW060036210926
848481LV00013B/3874